Medical Entomology for Students
FOURTH EDITION

Arthropod vectors of human infections, such as malaria, filariasis, West Nile virus, Lyme disease and typhus, are a continuing threat to human health. *Medical Entomology for Students* provides basic information on the recognition, biology, ecology and medical importance of the arthropods that affect human health. The fourth edition of this popular textbook is completely updated, and incorporates the latest strategies for controlling insects, ticks and mites. Extensive illustrations, with new colour photographs of some of the most important vectors and pests, will assist readers in recognizing arthropods such as mosquitoes, flies, and myiasis-causing larvae. The book contains a glossary of entomological and epidemiological terms, and a list of commonly used insecticides and their common names. Each chapter concludes with a list of suggested further reading. With a clear presentation and concise style, this text is specifically aimed at students of medical entomology, tropical medicine, parasitology, and pest control. It will also be essential reading for physicians, nurses, health officials and community health workers.

MIKE SERVICE is a world authority on medical entomology, with 50 years' experience of research and teaching. He is Emeritus Professor of Medical Entomology at the Liverpool School of Tropical Medicine, and has written over 200 research papers on medical entomology. In 1997 he was awarded the Sir Rickard Christophers medal by the Royal Society of Tropical Medicine and Hygiene, and in 2002 the Harry Hoogstraal Medal by the American Society of Tropical Medicine and Parasitology, for research on medical vectors.

Medical Entomology for Students

Fourth Edition

Mike Service

Emeritus Professor of Medical Entomology,
Liverpool School of Tropical Medicine, Liverpool, UK

CAMBRIDGE
UNIVERSITY PRESS

CAMBRIDGE UNIVERSITY PRESS
Cambridge, New York, Melbourne, Madrid, Cape Town, Singapore, São Paulo, Delhi

CAMBRIDGE UNIVERSITY PRESS
The Edinburgh Building, Cambridge, CB2 8RU, UK

www.cambridge.org
Information on this title: www.cambridge.org/9780521709286

First edition published by Chapman and Hall 1996
Second edition published by Cambridge University Press 2000
Third edition published 2004
Fourth edition published 2008
Third printing 2009

Printed in the United Kingdom at the University Press, Cambridge

A catalogue record for this publication is available from the British Library

ISBN 978-0-521-70928-6 paperback

earning Resources
Centre

13794647

Every effort has been made in preparing this book to provide accurate and
up-to-date information that is in accord with accepted standards and practice at
the time of publication. Nevertheless, the author, editors and publisher can make
no warranties that the information contained herein is totally free from error, not
least because clinical standards are constantly changing through research and
regulation. The author, editors and publisher therefore disclaim all liability for
direct of consequential damages resulting from the use of material contained in
this book. Readers are strongly advised to pay careful attention to information
provided by the manufacturer of any drugs or equipment that they plan to use.

To Wednesday yet again

Contents

Colour plate section appears between pages 150 and 151

Preface to the first edition

This is not intended as a reference book on medical entomology; those interested in such a book should consult *Medical Insects and Arachnids* (1993) edited by R. P. Lane and R. W. Crosskey (Chapman & Hall). The present book is aimed at students, whether they be physicians, nurses, health officials, community health workers or those studying for a masters' degree in parasitology or medical entomology. Its aim is to provide basic information on the recognition, biology and medical importance of arthropods and guidelines for their control. In a teaching book such as this it is always difficult to decide how much detail to include and what to omit; you cannot satisfy everyone. Nevertheless I have attempted to write a book to suit the needs of most students.

The reader should be selective. For example, I hope that most will find all, or most of, the information given in the chapters on fleas, lice, bed-bugs, scabies mites and flies relevant to their needs, but I would expect readers to be more selective with some other chapters, such as those on mosquitoes. These insects are undoubtedly the most important arthropod vectors; nevertheless some students may think I have included too much detail on certain aspects for their needs, and if so they can largely disregard such bits. I have also tried to be selective and avoid giving too many references at the end of each chapter, but with some vectors this has not been easy.

May 1995

Preface to the second edition

This new edition remains largely unchanged in respect of area covered from the first edition, and I have generally kept to the same style and format. As stated in the first edition this is a student textbook on medical entomology; those interested more in reference books should consult the publications listed at the end of this Preface. I have revised the text where necessary and rewritten many of the sections on control as these can quickly become outdated. A number of new figures are included, and some previous ones have been redrawn or modified. A few of the older references under the headings 'Further reading' have been omitted and several new ones added. Finally I have added a Glossary, mainly entomological, that I hope will help students to understand better some of the entomological terms commonly used.

February 1999

Preface to the third edition

The philosophy remains the same, to try and present clear and concise accounts of the most relevant information on the identification, life cycles and infections transmitted by arthropods of medical importance. The text has been completely revised and updated to include most recent vector and disease control strategies and new discoveries relating to the epidemiology of disease transmission. In 2000 some mosquito species formerly in the genus *Aedes* were transferred to the new genus *Ochlerotatus*, so that for example *Aedes togoi* is now called *Ochlerotatus togoi*. Although this change will inevitably lead to some confusion, I nevertheless have used it here because the name *Ochlerotatus* will now be found in most of the more recent scientific literature. Not to use this new generic name would create greater confusion! Some figures have been redrawn and new ones added, as have new tables and an appendix.

I have tried to help students during revisionary reading by placing in bold italics words relating to items, whether morphological (e.g. *antennae, capitulum*) or biological (e.g. *transovarial, reservoir hosts*), that are important in vector recognition or for understanding the role of vectors in disease transmission.

In addition to the Further reading at the end of each chapter, there is a Select bibliography of some key publications, mostly books on medical entomology, after the Glossary.

As before, readers, whether they be physicians, community health workers, health officials, nurses, or those specializing in medical entomology or parasitology, should be selective when reading the various chapters in this book, and focus on facts and issues that are most relevant to their needs or studies. Good luck!

October 2003

Preface to the fourth edition

The layout and approach in this edition remains the same as in previous editions but the text has been revised and updated, with the biggest changes made to vector control procedures. Problems can, however, arise when advocating the use of named insecticides for control because some insecticides (acaricides) may be banned in some countries yet not in others. In general I have left it to the reader to find out what chemicals are allowed in his or her area of operation.

Two new transmission cycles on West Nile and Japanese encephalitis viruses have been incorporated in Chapter 3. However, the biggest change to this edition is the inclusion of 24 colour illustrations of some of the more important vectors and pest species. This it is hoped will help readers have a better appreciation of what various arthropods actually look like – and some may even find the photo of a *Sabethes* mosquito beautiful! Another change has been to return to the older classification of mosquitoes and place all relevant species in the genus *Aedes*; that is, *Ochlerotatus* is not recognized here as a genus. Apart from other considerations this will simplify matters for readers.

It has been difficult to decide on the amount of detail and information that should be included in a book that is essentially a primer on medical entomology. Readership is targeted at a widely diverse audience, including community health workers, health officials working in public health programmes, nurses and physicians as well as those specializing in medical entomology or parasitology. However, as before I suggest that readers should be selective in deciding the amount of detail in the book they need to know, and thus focus on the most relevant issues.

August 2007

Acknowledgements

I would like to thank the following colleagues and friends, arranged in alphabetical order, for their advice and help with various questions I had on their specialist subjects: R. W. Ashford, J. H. Bryan, R. W. Crosskey, C. Curtis, J. B. Davies, M. J. Donnelly, M. J. R. Hall, R. E. Harbach, N. C. Hinkle, A. M. Jordan, A. Krueger, J. W. McGarry, D. H. Molyneux, A. Pont, R. J. Post, R. Robbins and A. Walker.

I would like to acknowledge the generosity of Killgerm Group Ltd UK for covering the cost of publishing the coloured figures in the book.

I am indebted to the Trustees of the Natural History Museum, London, for permission to reproduce Figures 4.2 and 4.4a, drawn by R. W. Crosskey, and to the Museum for permission to reproduce, in a modified form, Figures 17.1 (male tick) and 19.3. I am also indebted to Blackwell Publishing, Oxford, for permission to reproduce Figures 1.14 and 11.5, drawn by Miss M. A. Johnson.

Finally, I am indebted to Hugh Brazier, copy-editor for Cambridge University Press, for his diligence in picking up many minor errors and a few not so minor ones and for suggesting how certain parts of the text could be improved, all to the benefit of the readers. Undoubtedly Hugh has been the best copy-editor I have ever worked with.

1

Introduction to mosquitoes (Culicidae)

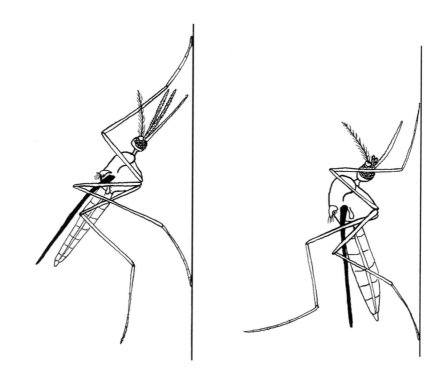

There are some 3400 species of mosquitoes, which are traditionally placed in 42 genera, all contained in the family Culicidae. Some mosquito experts recognize many more genera. For example, some mosquitoes in the genus *Aedes* have been transferred to the genera *Ochlerotatus* and *Stegomyia*, but these genera are not used in this book as they are regarded as subgenera.

The Culicidae is divided into three subfamilies: Toxorhynchitinae, Anophelinae (anophelines) and Culicinae (culicines). Mosquitoes have a worldwide distribution, occurring throughout the tropical and temperate regions and northwards into the Arctic Circle. The only areas from which they are absent are Antarctica, and a few islands. They are found at elevations of 3500 m and down mines to depths of 1250 m below sea level.

The most important pest and vector species belong to the genera *Anopheles*, *Culex*, *Aedes*, *Psorophora*, *Mansonia*, *Haemagogus* and *Sabethes*. *Anopheles* species, as well as transmitting malaria, are vectors of filariasis (*Wuchereria bancrofti*, *Brugia malayi* and *Brugia timori*) and a few arboviruses. Certain *Culex* species transmit *Wuchereria bancrofti* and a variety of arboviruses. *Aedes* species are important vectors of yellow fever, dengue, West Nile virus and many other arboviruses, and in a few restricted areas they are also vectors of *Wuchereria bancrofti* and *Brugia malayi*. *Mansonia* species transmit *Brugia malayi* and sometimes *Wuchereria bancrofti* and a few arboviruses. *Haemagogus* and *Sabethes* mosquitoes are vectors of yellow fever and a few other arboviruses in Central and South America, while the genus *Psorophora* contains some troublesome pest species in North and South America, as well as a few that transmit arboviruses.

Many species which are not vectors can nevertheless be troublesome because of the serious biting nuisance they cause.

1.1   External morphology

Mosquitoes possess only one pair of functional wings, the fore-wings. The hind-wings are represented by a pair of small, knob-like halteres. Mosquitoes are distinguished from other flies of a somewhat similar shape and size by: (1) the possession of a conspicuous forward-projecting proboscis; (2) the presence of numerous appressed scales on the thorax, legs, abdomen and wing veins; and (3) a fringe of scales along the posterior margin of the wings.

Mosquitoes are slender and relatively small insects, usually measuring about 3–6 mm in length. Some species, however, can be as small as 2 mm while others may be as long as 19 mm. The body is distinctly divided into a head, thorax and abdomen.

The head has a conspicuous pair of kidney-shaped compound eyes. Between the eyes arises a pair of filamentous and segmented antennae. In females the antennae have whorls of short hairs (that is pilose antennae), but in males, with a few exceptions in genera of no medical importance, the antennae have many long hairs giving them a feathery or plumose

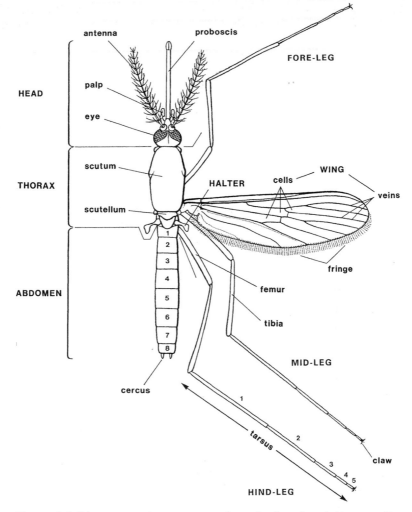

Figure 1.1 Diagrammatic representation of a female adult mosquito.

appearance. Mosquitoes can thus be conveniently sexed by examination of their ***antennae***: individuals with feathery antennae are males, whereas those with only short and rather inconspicuous antennal hairs are females (Figs. 1.1, 1.13). Just below the antennae is a pair of ***palps***, which may be long or short and dilated or pointed at their tips, depending on the sex of the adults and whether they are anophelines or culicines (Fig. 1.13). Arising between the palps is the single long proboscis, which contains the piercing mouthparts. In mosquitoes the proboscis characteristically projects forwards (Fig. 1.1).

The thorax is covered, dorsally and laterally, with scales which may be dull or shiny, white, brown, black or almost any colour. It is the

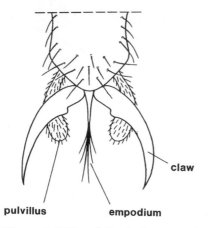

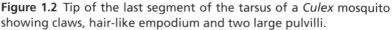

Figure 1.2 Tip of the last segment of the tarsus of a *Culex* mosquito showing claws, hair-like empodium and two large pulvilli.

arrangement of black and white, or coloured, scales on the dorsal surface of the thorax that gives many species, especially *Aedes* mosquitoes, their distinctive patterns (Fig. 3.3).

The wings are long and relatively narrow, and the number and arrangement of the wing veins is virtually the same for all mosquito species (Fig. 1.1). The veins are covered with scales which are usually brown, black, white or yellowish, but more brightly coloured scales may occasionally be present. The shape of the scales and the pattern they form differs considerably between both genera and species of mosquitoes. Scales also project as a fringe along the posterior border of the wings. In life the wings of resting mosquitoes are placed across each other over the abdomen in the fashion of a closed pair of scissors. The legs are long and slender and are covered with scales which are usually brown, black or white and may be arranged in patterns, often in the form of rings (Fig. 3.4b). The tarsus usually terminates in a pair of toothed or simple claws. Some genera, such as *Culex*, have a pair of small fleshy pulvilli (Fig. 1.2) between the claws.

The abdomen is composed of 10 segments but only the first seven or eight are visible. Mosquitoes in the subfamily Culicinae usually have the abdomen covered dorsally and ventrally with mostly brown, blackish or whitish scales. In the Anophelinae, however, the abdomen is almost, or entirely, devoid of scales. The last abdominal segment of a female mosquito terminates in a pair of small finger-like cerci, whereas in the male there is a pair of prominent claspers, comprising part of the male external genitalia.

In unfed mosquitoes the abdomen is thin and slender, but after females have bitten a suitable host and taken a blood-meal (only females bite) the abdomen becomes greatly distended and resembles an oval red balloon.

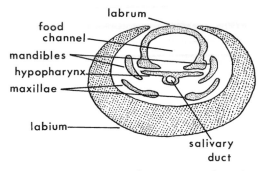

Figure 1.3 Diagram of a cross-section through the proboscis of a mosquito, showing components of the mouthparts and food channel.

When the abdomen is full of developing eggs it is also dilated, but is whitish and not red in appearance.

1.1.1 Mouthparts and salivary glands

The mouthparts are collectively known as the *proboscis*. In mosquitoes the proboscis is long and projects conspicuously forwards in both sexes – although males do not bite. The largest component of the mouthparts is the long and flexible gutter-shaped labium, which terminates in a pair of small flap-like structures called labella. In cross-section the labium is seen to almost encircle all other components of the mouthparts (Fig. 1.3) and serves as a protective sheath. The individual components are held close together in life and only become partially separated during blood-feeding, or when they are teased apart for examination as illustrated in Figure 1.4.

The uppermost structure, the labrum, is slender, pointed and grooved along its ventral surface. In between this 'upper roof' (labrum) and 'lower gutter' (labium) are five needle-like structures, namely a lower pair of toothed maxillae, an upper pair of mandibles, which usually lack teeth (although in *Anopheles* they are very finely toothed), and finally a single untoothed, hollow stylet called the hypopharynx. When a female mosquito bites a host the labella, at the tip of the fleshy labium, are placed on the skin and the labium, which cannot pierce the skin, curves backwards. This allows the paired mandibles, paired maxillae, labrum and hypopharynx to penetrate the host's skin. Saliva from a pair of trilobed salivary glands, situated ventrally in the anterior part of the thorax, is pumped down the hypopharynx. *Saliva* contains antihaemostatic enzymes that produce haematomas in the skin and facilitate the uptake of blood. Saliva also contains anticoagulants which prevent blood from clotting and obstructing the mouthparts as it is sucked up, and anaesthetic substances that help reduce the pain inflicted by the mosquito's bite, so reducing the host's defensive reactions.

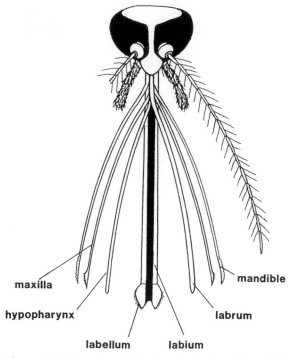

maxilla

hypopharynx

mandible

labrum

labellum   labium

Figure 1.4 Diagram of the head of a female cuclicine mosquito, showing the components of the mouthparts spread out from the labium.

Although male mosquitoes have a proboscis, the maxillae and mandibles are usually reduced in size or the mandibles are absent, so males cannot bite.

1.2   Life cycle

1.2.1   Blood-feeding and the gonotrophic cycle

Most mosquitoes mate shortly after emergence from the pupa. Sperm from a male enters the spermotheca of a female and usually serves to fertilize all eggs laid during her lifetime; thus only one mating and insemination per female is required. With a few exceptions, a female mosquito must bite a host and take a blood-meal to obtain the necessary nutrients for the development of her eggs. This is the normal procedure and is referred to as *anautogenous* development. A few species, however, can develop the first batch of eggs without a blood-meal, and more rarely subsequent batches. This process is called *autogenous* development. The speed of digestion of the blood-meal depends on temperature. In most tropical species it takes only 2–3 days, but in colder, temperate countries blood digestion may take as long as 7–14 days.

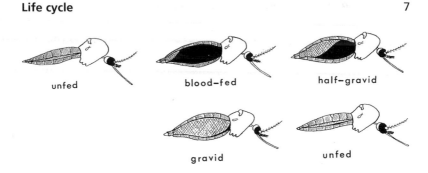

unfed   blood-fed   half-gravid

gravid   unfed

Figure 1.5 Diagrammatic representation of the gonotrophic cycle of a female mosquito. Each cycle begins with an unfed adult, which passes through a blood-fed, half-gravid and gravid condition. After oviposition the female is again unfed and seeks another blood-meal.

After a blood-meal the mosquito's abdomen is dilated and bright red in colour, but some hours later the abdomen becomes a much darker red. As the blood is digested and the white eggs in the ovaries enlarge, the abdomen becomes whitish posteriorly and dark reddish anteriorly. This condition represents a mid-point in blood digestion and ovarian development, and the mosquito is referred to as being half-gravid (Fig. 1.5). Eventually all blood is digested and the abdomen becomes dilated and whitish due to the formation of fully developed eggs (Fig. 1.5). The female is now said to be gravid and she searches for suitable larval habitats in which to lay her eggs. After oviposition the female mosquito takes another blood-meal and after 2–3 days (in the tropics) a further batch of eggs is matured. This process of blood-feeding and egg maturation, followed by oviposition, is repeated several times throughout the female's life and is referred to as the *gonotrophic cycle*.

Male mosquitoes cannot bite but feed on the nectar of flowers and other naturally occurring sugary secretions. Males are consequently unable to transmit any diseases. Sugar feeding is not, however, restricted to males: females may also feed on sugary substances to obtain energy for flight and dispersal, but only in a few species (the autogenous ones) is this type of food sufficient for egg development.

1.2.2 Oviposition and biology of the eggs

Depending on the species, female mosquitoes lay about 30–300 eggs at any one oviposition. Eggs are brown or blackish and 1 mm or less in length. In many Culicinae they are elongate or approximately ovoid in shape, but eggs of *Mansonia* are drawn out into a terminal filament (Fig. 3.8). In the Anophelinae eggs are usually boat-shaped (Fig. 1.8). Many mosquitoes, such as species of *Anopheles* and *Culex*, lay their eggs directly on the water surface. In *Anopheles* the eggs are laid singly and float on the water, whereas those of *Culex* are laid vertically in several rows held together by surface tension to form an *egg raft* which floats on the water (Fig. 1.15). *Mansonia* species lay their eggs in a sticky mass that is glued to the underside of floating plants. None of the eggs of these mosquitoes can survive desiccation and consequently they die if they become dry. In the tropics eggs hatch within 2–3 days, but in cooler temperate countries they may not hatch until after 7–14 days, or longer.

Other mosquitoes, such as those belonging to the genera *Aedes*, *Psorophora* and *Haemagogus*, do not lay eggs on the water surface. Instead they deposit them just above the water line on damp substrates, such as mud and leaf litter, or on the inside walls of tree-holes and clay water-storage pots. Eggs of these genera can withstand desiccation, especially those of *Aedes* and *Psorophora*, which can remain dry for months or even years but still remain viable and hatch when soaked in water. Because their eggs are laid above the water line of larval habitats it may be many weeks or months before they become flooded with water and can hatch. However, even when flooded, hatching may extend over long periods because the eggs hatch in instalments. Moreover, eggs of *Aedes* and *Psorophora* may require repeated immersions in water followed by short periods of desiccation before they will hatch. *Aedes* and *Psorophora* eggs may also enter a state of *diapause*, that is not hatching until some specific environmental stimulus such as a change in daylength and/or temperature breaks diapause and the eggs hatch. In temperate regions many *Aedes* and *Psorophora* species overwinter as diapausing eggs.

1.2.3   Larval biology

Mosquito larvae are distinguished from most other aquatic insects by being legless and having a bulbous thorax that is wider than both the head and the abdomen. There are four active larval *instars*. All mosquito larvae require water in which to develop; no mosquito has larvae that can withstand desiccation, although they may be able to survive short periods, for example, in wet mud.

Larvae have a well-developed head bearing a pair of antennae and a pair of compound eyes. Prominent mouthbrushes are present in most species and serve to sweep water containing minute food particles into the mouth. The thorax is roundish and has unbranched and branched hairs, which are usually long and conspicuous. The 10-segmented abdomen has nine visible segments, most of which have unbranched or branched hairs (Figs. 1.9, 1.16). The last segment, which differs in shape from the preceding eight segments, has two paired groups of long hairs forming the caudal setae, and a larger fan-like group comprising the *ventral brush* (Figs. 1.10, 1.16). This last segment ends in two pairs of transparent, sausage-shaped anal papillae, which although often called gills are concerned not with respiration but with osmoregulation.

Mosquito larvae, with the exception of *Mansonia* and *Coquillettidia* species (and a few other species), must come to the water surface to breathe. Atmospheric air is taken in through a pair of spiracles situated dorsally on the ninth abdominal segment. In the subfamilies Toxorhynchitinae and Culicinae these spiracles are situated at the end of a single dark-coloured and heavily sclerotized tube termed the **siphon** (Fig. 1.16). *Mansonia* and *Coquillettidia* larvae possess a specialized siphon that is more or less conical, pointed at the tip and supplied with prehensile hairs and serrated cutting structures (Fig. 3.9). These enable the siphon to be inserted into the roots or stems of aquatic plants, from which oxygen for larval respiration is obtained. In contrast, larvae of the Anophelinae do not have a siphon (Figs. 1.10, 1.13).

Mosquito larvae feed on yeasts, bacteria, protozoans and numerous other micro-organisms, as well as on decaying plant and animal material found in the water. Some, such as *Anopheles* species, are surface-feeders, whereas many others browse over the bottoms of habitats. A few mosquitoes are carnivorous or cannibalistic. There are four larval instars and in tropical countries larval development, that is the time from egg hatching to pupation, can be as short as 5–7 days, but many species require about 7–14 days. In temperate areas the larval period may last several weeks or months, and several species overwinter as larvae.

1.2.4   Larval habitats

Mosquito larval habitats vary from large and usually permanent collections of water, such as freshwater swamps, marshes, ricefields and borrow pits, to smaller collections of temporary water such as pools, puddles, water-filled car tracks, ditches, drains and gulleys. A great variety of 'natural container-habitats' also provide breeding places, such as water-filled tree-holes, rock-pools, water-filled bamboo stumps, bromeliads, pitcher plants, leaf axils in banana, pineapple and other plants, water-filled split coconut husks and snail shells. Larvae also occur in wells and 'man-made container-habitats', such as clay pots, water-storage jars, tin cans, discarded kitchen utensils and motor vehicle tyres. Some species prefer shaded larval habitats whereas others like sunlit habitats. Many species cannot survive in water polluted with organic debris whereas others can breed prolifically in water contaminated with excreta or rotting vegetation. A few mosquitoes breed almost exclusively in brackish or salt waters, such as saltwater marshes and mangrove swamps, and are consequently restricted to mostly coastal areas. Some species are less specific in their requirements and can tolerate a wide range of different types of breeding place.

Almost any collection of permanent or temporary water can be a mosquito larval habitat, but larvae are usually absent from large expanses of uninterrupted water such as lakes, especially if they have large numbers of fish and other predators. They are also usually absent from large rivers

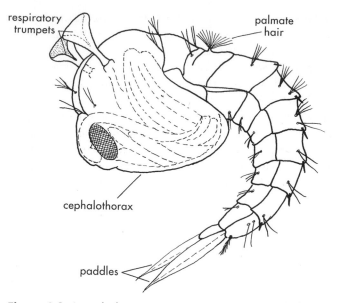

Figure 1.6 *Anopheles* pupa.

and fast-flowing waters, but they may occur in marshy areas and isolated pools and puddles formed at the edges of flowing water.

1.2.5   Pupal biology

All mosquito pupae are aquatic and comma-shaped. The head and thorax are combined to form the *cephalothorax*, which dorsally has a pair of respiratory *trumpets* (Fig. 1.6). The abdomen is 10-segmented, although only eight segments are visible. Each segment has numerous short hairs and the last segment terminates in a pair of oval and flattened structures termed *paddles* (Figs. 1.11, 1.18). Some of the developing structures of the adult mosquito can be seen through the integument of the cephalothorax, the most conspicuous features being a pair of dark compound eyes, folded wings, legs and the proboscis (Fig. 1.6).

Pupae do not feed but spend most of their time at the water surface taking in air through the respiratory trumpets. If disturbed they swim up and down in the water in a jerky fashion.

Pupae of *Mansonia* and *Coquillettidia* differ in that they have relatively long breathing trumpets, which are modified to enable them to pierce aquatic vegetation and obtain their oxygen in a similar fashion to the larvae (Fig. 3.9). As a consequence their pupae remain submerged and rarely come to the water surface.

In the tropics the pupal period lasts only 2–3 days but in cooler temperate regions pupal development may take 9–12 days, or longer. At the end of pupal life the skin on the dorsal surface of the cephalothorax splits, and the adult mosquito struggles out.

1.2.6   Adult biology and behaviour

As already mentioned (p. 6), females of most mosquito species require a blood-meal before the eggs can develop, and this is taken either before or more usually after mating. Many species bite humans to obtain their blood-meals and a few feed on humans in preference to other animals. However, others prefer feeding on non-human hosts and many species never bite people. Species that usually feed on humans are said to be *anthropophagic* in their feeding habits, whereas those that feed mainly on other animals are called *zoophagic*. Mosquitoes that feed on birds are sometimes called *ornithophagic* instead of zoophagic. Females are attracted to hosts by various stimuli emanating from their breath or sweat, such as carbon dioxide, lactic acid, octenol, as well as body odours and warmth. Vision usually plays only a minor role in host orientation. Some species feed more or less indiscriminately at any time of the day or night; others are mainly diurnal or nocturnal in their biting habits.

A few species of mosquitoes frequently enter houses to feed and are said to be *endophagic* in their feeding habits, whereas those that bite their hosts outside houses are called *exophagic*. After having bitten humans, or some other hosts, either inside or outside houses, mosquitoes seek resting places in which to shelter during digestion of their blood-meals. Some species rest inside houses during the time required for blood digestion and maturation of the ovaries and are called *endophilic*. In contrast mosquitoes that rest outdoors are termed *exophilic*. Female adults of *Aedes aegypti* (a vector of yellow fever), for example, are usually anthropophagic, exophagic and exophilic, whereas adults of *Anopheles gambiae* (African malaria vector) are mainly anthropophagic, endophagic and endophilic. Few mosquitoes, however, are entirely anthropophagic or zoophagic, endophagic or exophagic, endophilic or exophilic. Instead, most show various degrees of these behavioural patterns; in other words all these terms are relative. The feeding behaviour of a species may also change. For example, in certain areas and at certain seasons a species may bite people predominantly (anthropophagic) inside houses (endophagic) and remain in houses afterwards (endophilic), whereas at other times, especially if there are few people but many animals in the area, the species may become predominantly zoophagic, and also exophagic and exophilic. Many species are less adaptable in their feeding behaviour and will never rest inside houses or enter them to feed on people.

The *biting behaviour* of female mosquitoes may be very important in the epidemiology of disease transmission. Mosquitoes feeding on people predominantly out of doors and late at night will not bite many young children, because they will be indoors and asleep at this time. Consequently young children will be less likely to be infected with any diseases that these mosquitoes might transmit. During hot and dry seasons substantial numbers of people may sleep out of doors and as a consequence be bitten more frequently by exophagic mosquitoes. Some mosquitoes bite predominantly within forests or wooded areas, so people will only get bitten when they visit these places. Clearly the behaviour of both people and mosquitoes may be relevant in disease transmission.

The resting behaviour of adult mosquitoes may be important in planning control measures. In several malaria control campaigns the interior surfaces of houses, such as walls and ceilings, are sprayed with residual insecticides to kill adult mosquitoes resting on them. This approach will, of course, only be effective in controlling malaria if the mosquito vectors are endophilic (see p. 47).

Most mosquitoes probably disperse only a few hundred metres from their emergence sites, and in control programmes and epidemiological studies it is usually safe to say that mosquitoes will not fly further than 2 km. There are, however, records of mosquitoes being found up to 100 km or more from their breeding places, but such dispersal is nearly always wind-assisted. Mosquitoes may get transported long distances in aeroplanes, and sometimes this causes disease outbreaks, such as 'airport malaria'.

In tropical countries adult female mosquitoes probably live *on average* 1–2 weeks, whereas in temperate countries adult longevity is likely to be 3–4 weeks. Species that hibernate or aestivate live much longer: for example in Europe some fertilized female *Culex pipiens* survive in hibernation from August until May. Adult males usually have a shorter life span than females.

1.3   Classification of mosquitoes

1.3.1   Subfamily Toxorhynchitinae

Subfamily Toxorhynchitinae comprises a single genus, *Toxorhynchites*, which contains about 94 species that are mainly tropical, although a few species occur in North America, south-eastern Russia and Japan.

Adults are large (19 mm long, 24 mm wingspan) and colourful, being metallic bluish or greenish with black, white or red tufts of hair-like scales projecting from the posterior abdominal segments. Adults are easily recognized by having a proboscis that is curved backwards in both sexes (Fig. 1.7) and that is incapable of piercing the skin. Consequently, since neither sex can bite, they are of no medical importance. Their larvae are also large (12–18 mm long), often dark reddish and, like those of the Culicinae, have a siphon. They are predaceous on larvae of other mosquitoes and on their own kind. They have occasionally been introduced into areas in the hope that their voracious larvae will help reduce the numbers of

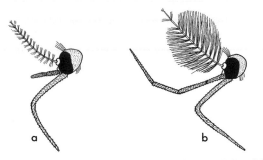

Figure 1.7 Heads of *Toxorhynchites* adults: (a) female; (b) male.

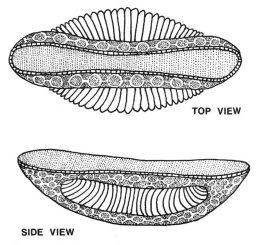

TOP VIEW

SIDE VIEW

Figure 1.8 *Anopheles* eggs.

pest mosquitoes. Larvae are found mainly in container-habitats, such as tree-holes and bamboo stumps, tin cans and water-storage pots.

1.3.2   Subfamily Anophelinae

Of the three genera included in the subfamily Anophelinae only the genus *Anopheles* (about 484 species) is of medical importance, for instance as malaria vectors. The following characters serve to separate *Anopheles* mosquitoes from those in any of the other mosquito genera.

Anopheline eggs

Eggs are laid singly on the water surface. In most species they are typically boat-shaped, and laterally have a pair of air-filled sacs called *floats* (Fig. 1.8). Anopheline eggs are unable to withstand desiccation.

Anopheline larvae

Larvae lack a siphon and lie *parallel* to the water surface, not subtended at an angle as are the culicines. They are surface-feeders and so spend most of their time at the water surface. Examination under a microscope shows that the abdomen has small, brown, sclerotized plates, called *tergal plates*, on the dorsal surface of abdominal segments 1–8; there may also be 1–3 small accessory plates behind the main tergal plate. In addition most or all of these segments have a pair of well-developed *palmate hairs*, sometimes called float hairs (Figs. 1.9, 1.10). These abdominal palmate hairs and a single pair on the thorax come into contact with the water surface and aid in keeping larvae parallel to the surface. Laterally on each side of segment 8 (8 and 9 are combined) there is a sclerotized structure with teeth called the *pecten*. All these structures identify larvae as belonging to the genus *Anopheles*.

Anopheline pupae

The respiratory trumpets of anopheline pupae are short and broad distally, thus appearing conical (Figs. 1.6, 1.11a), whereas in most culicines the trumpets are narrower and more cylindrical. The most reliable character for identifying anopheline pupae is the presence of short, peg-like spines situated laterally near the distal margins of abdominal segments 2–7 or 3–7 (Fig. 1.11b); in culicines there are no such spines.

Anopheline adults

Adult *Anopheles* usually rest with their bodies at an *angle* to the surface, that is with the proboscis and abdomen in a straight line, or 'head down bottom up' (Fig. 1.13). In some species they rest at almost right angles to the surface, whereas in others such as the Indian mosquito *Anopheles culicifacies* the angle is much smaller. This is a very useful character, allowing adults resting in houses and elsewhere to be readily identified as *Anopheles*.

Most, but not all, *Anopheles* mosquitoes have the dark (usually black) and pale (usually white or yellowish) scales on the wing veins arranged in '*blocks*' or specific areas (Fig. 1.12) forming a distinctive spotted pattern which differs according to species. A few species, however, such as the European *Anopheles claviger*, have the veins covered more or less uniformly with dark (often brown) scales.

The most reliable way to distinguish between adult *Anopheles* and Culicinae is by examination of their *heads*. The first procedure is to determine the sex of the adults: female mosquitoes have non-plumose antennae whereas males have plumose antennae. If the adults are females and also *Anopheles* then the *palps* will be about as long as the proboscis and usually lie closely alongside it (Fig. 1.13). The palps are usually blackish with broad or narrow rings of pale scales, especially on the apical half. In male *Anopheles* the *palps* are also about as long as the proboscis but are distinctly swollen at

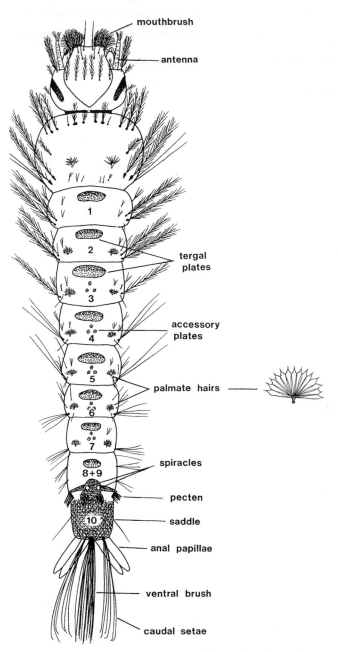

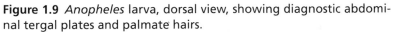

Figure 1.9 *Anopheles* larva, dorsal view, showing diagnostic abdominal tergal plates and palmate hairs.

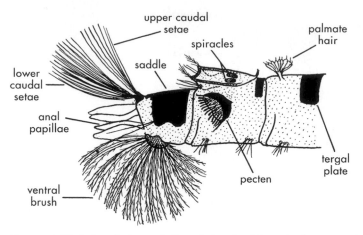

Figure 1.10 Lateral view of the abdominal terminal segments of an *Anopheles* larva.

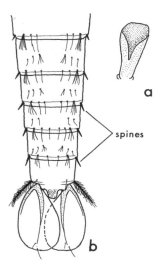

Figure 1.11 *Anopheles* pupa: (a) short and broad respiratory trumpet; (b) part of abdomen showing diagnostic spines.

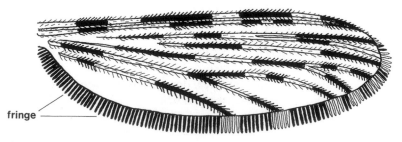

Figure 1.12 *Anopheles* wing, showing dark and pale scales arranged in 'blocks'.

ANOPHELINES | CULICINES

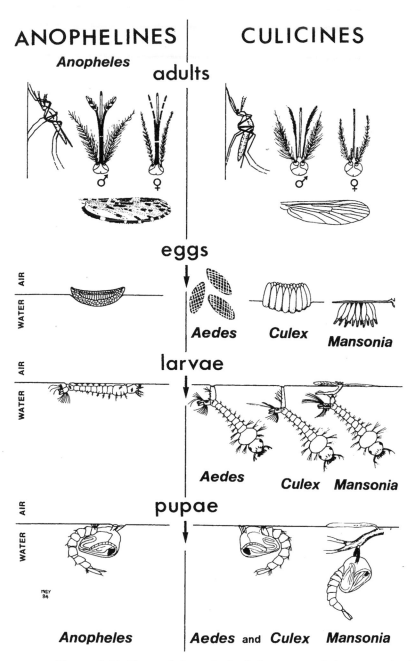

Figure 1.13 Chart of the principal characters of the stages in the life cycle that distinguish anopheline from culicine mosquitoes.

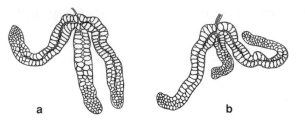

a   b

Figure 1.14 Salivary glands of adult mosquitoes: (a) culicine; (b) *Anopheles*. (Courtesy of Miss M. A. Johnson, and Blackwell Publishing, Oxford, publishers of *Entomology for Students of Medicine* (1962) by R. M. Gordon and M. M. J. Lavoipierre.)

the ends and are said to be clubbed (Fig. 1.13); they may also have rings of pale scales apically.

Other differences are that in *Anopheles* there is only a single spermotheca in females, and in both sexes the middle lobe of the salivary glands is considerably shorter than the two outer lobes (Fig. 1.14).

Principal characters for separating the various stages in the life cycles of anopheline and culicine mosquitoes are given in Table 1.1.

1.3.3   Subfamily Culicinae

There are some 2750 species in the large subfamily Culicinae, belonging to 38 genera (some taxonomists recognize several more genera). The most important medically are the genera *Aedes*, *Culex*, *Mansonia*, *Haemagogus*, *Sabethes* and *Psorophora*. The following characters separate the Culicinae from *Anopheles* mosquitoes. Methods for distinguishing the more important genera within the Culicinae are given in Chapter 3.

Culicine eggs

Culicine eggs never have floats. They are laid either singly (e.g. *Aedes*) or in the form of **egg rafts** that float on the water surface (e.g. *Culex* and *Coquillettidia*), or are deposited as sticky masses glued to the underside of floating vegetation (e.g. *Mansonia*) (Fig. 1.15).

Culicine larvae

All culicine larvae possess a **siphon** (Fig. 1.16), which may be long or short. They hang upside down at an angle from the water surface when they are getting air (Fig. 1.13), except for *Mansonia* and *Coquillettidia* larvae, which insert their specialized siphons into aquatic plants and remain submerged (Fig. 3.9). There are no abdominal palmate hairs or tergal plates on culicine larvae.

Table 1.1. *Principal characters distinguishing anopheline and culicine mosquitoes*

Stage	Anophelinae	Culicinae
Eggs	Laid singly, possess floats	Laid singly or in egg rafts or masses. Never possess floats
Larvae	Never have a siphon. Lie parallel to water surface. Have abdominal palmate hairs and tergal plates	All larvae have a short or long siphon. Subtend an angle from the water surface. No palmate hairs or tergal plates
Pupae	Breathing trumpets short and broad apically. Short peg-like spines on abdominal segments 2–7 or 3–7	Breathing trumpets short or long, opening not broad. No spines on abdominal segments 2–7
Adults (both sexes)	Rest at an angle to any surface. In most species dark and pale scales on wing veins arranged in distinct 'blocks'	Rest with body more or less parallel to the surface. Scales on wing veins not arranged in 'blocks'; scales frequently all brown or blackish, or a mixture of pale and dark scales scattered on veins
Adult females (non-plumose antennae)	Palps about as long as proboscis	Palps much shorter than proboscis
Adult males (plumose antennae)	Palps about as long as proboscis and swollen at ends	Palps about as long as proboscis but never swollen at ends; palps may be hairy distally

Culicine pupae

The length of the respiratory trumpets in culicine pupae is variable, but they are generally longer, more cylindrical and have narrower openings (Fig. 1.17) than in *Anopheles*. Abdominal segments 2–7 lack peg-like spines, although they have numerous setae (Fig. 1.18).

Culicine adults

Culicine adults rest with the thorax and abdomen more or less **parallel** to the surface (Fig. 1.13). The scales covering the wing veins are commonly

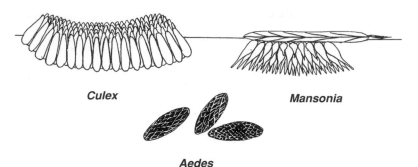

Culex   *Mansonia*

Aedes

Figure 1.15 Mosquito eggs. *Culex* egg raft that floats on the water surface, *Mansonia* eggs glued to the undersurface of floating aquatic vegetation, and individual *Aedes* eggs that are deposited on damp surfaces.

uniformly brown or black. Sometimes, however, there are contrasting dark and pale scales but they are not arranged in distinctive areas or 'blocks', as found in many *Anopheles* adults.

The most reliable method for identifying the Culicinae is to examine their **heads**. In females (which have non-plumose antennae) the **palps** are shorter than the proboscis. In males (which have plumose antennae) the **palps** are about as long as the proboscis but are not swollen distally and hence do not appear clubbed (Fig. 1.13). However, the palps may be turned upwards distally, and in many species they are covered with long hairs so that superficially they can appear to be somewhat swollen apically, but more careful examination shows that the palps in male culicines are not clubbed as they are in *Anopheles*.

Other differences separating the Culicinae from *Anopheles* are that in culicines there are two or three spermathecae in females, but just one in anophelines. Also in culicines the middle lobe of the salivary glands is about as long as the other two, whereas in anophelines it is shorter (Fig. 1.14).

1.4   Medical importance

Although in many temperate countries mosquitoes may be of little or no importance as disease vectors they can, nevertheless, cause considerable annoyance because of their bites. The greatest numbers of mosquitoes are found in the northern areas of the temperate regions, especially near or within the Arctic Circle, where the numbers biting can be so great at certain times of the year as to make almost any outdoor activity impossible. Because of their elongated mouthparts female mosquitoes have little difficulty in biting through clothing such as socks, shirts, blouses, trousers and woollen garments, but clothing with a much closer weave of material may prevent this.

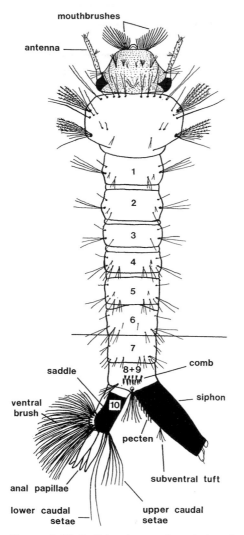

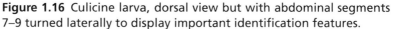

Figure 1.16 Culicine larva, dorsal view but with abdominal segments 7–9 turned laterally to display important identification features.

Mosquitoes are important as vectors of malaria, various forms of filariasis and numerous arboviruses, the best known being dengue, yellow fever and West Nile virus. Their role in the transmission of these diseases is discussed in Chapters 2 and 3.

1.5   Mosquito control

Control measures which are directed against specific vectors, such as anopheline malaria vectors, and *Aedes aegypti* and *Culex quinquefasciatus*,

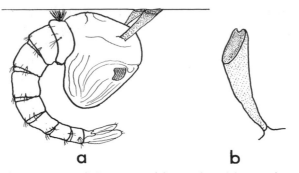

a   **b**

Figure 1.17 Culicine pupa: (a) pupal position at the water surface; (b) one of the two elongated and relatively narrow respiratory trumpets.

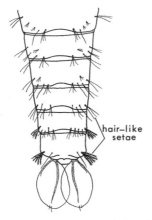

hair–like
setae

Figure 1.18 Part of the abdomen of a culicine pupa, showing hair-like setae (note absence of spines as found on anopheline pupae).

are described in more detail in Chapters 2 and 3; only the more basic principles of control are outlined here.

Control measures can be directed at either the immature aquatic stages or the adults, or at both stages simultaneously.

1.5.1   Control directed at the immature stages
Biological control

Although often termed naturalistic control, there is little that is natural about biological control. Either the numbers of predators, parasites or pathogens in habitats must be greatly increased to obtain worthwhile control, or they have to be introduced into habitats from which they were originally absent; such environmental manipulations are not natural. Biological control of mosquitoes was very popular during the early twentieth century, but with the availability of chemicals such as the organochlorines and organophosphates it was replaced by insecticidal control. However, because of insecticide resistance and greater awareness of environmental contamination there has been renewed interest in biological (biocontrol) methods. They are, however, usually more difficult to implement and maintain than insecticidal methods. Moreover, with predators it is unlikely that they will prey exclusively on mosquito larvae and pupae but will also eat harmless or even beneficial insects. Finally, biological control does not lead to rapid control. It takes some days, or more often weeks, before pest or vector mosquito populations are substantially reduced in size.

Predators Larvivorous fish are the most widely used biological control agents, the most common being the top minnow or mosquitofish (*Gambusia affinis*). This is a warm-water fish originally native to the southern USA and northern Mexico but which has been introduced to over 70 countries worldwide, including the Pacific islands, Europe, the Middle East, India, South-east Asia and Africa, in attempts to control mosquito larvae. They are aggressive fish which have sometimes destroyed indigenous species; consequently they should not now be introduced into new areas. Another commonly used fish is the South American guppy (*Poecilia reticulata*), which is not so voracious as *G. affinis* but can better tolerate low levels of organic pollution and is more heat-tolerant. There are numerous other fish that have been used to eat mosquito larvae, such as carp (e.g. *Cyprinus carpio*) in Chinese ricefields, an edible catfish (*Clarias fuscus*) in water-storage tanks in Myanmar to control *Aedes aegypti*, and *Oreochromis* (= *Tilapia*) species in Africa and *Aplocheilus* species in Europe and Asia.

Predatory fish, such as *Aphanius dispar* and *Fundulus* species, breed in saline waters and can therefore be introduced into saltwater habitats. Fish are unsuitable for controlling mosquitoes in small water containers and in pools and puddles that rapidly dry out. However, some fish, such as species of *Nothobranchius* and *Cynolebias*, which are the so-called instant or annual fish, have drought-resistant eggs, and these are more suitable for introducing into small temporary habitats that repeatedly dry out.

Although fish have sometimes greatly reduced the numbers of larvae in certain habitats, such as borrow pits, ponds, wells and ricefields, other than in parts of India and China they they have rarely proved effective in reducing mosquito populations over relatively large areas. Nor is there usually much convincing evidence that they have significantly decreased the incidence of mosquito-borne diseases – but see page 46 in relation to malaria.

Other predators of mosquito larvae include tadpoles of frogs and toads and various aquatic insect larvae, but these have rarely proved effective as control agents. Over the last 10 years there has been increasing interest in predaceous *Mesocyclops* species (copepods) in controlling larvae in water containers such as tyres, but the impact is usually very localized. A few mosquitoes have predaceous larvae, for example *Toxorhynchites* species. These have been introduced into container-habitats in certain areas (e.g. Fiji, Samoa and Hawaii) to control other mosquito larvae, but results have not been very encouraging.

Pathogens and parasites   There are numerous pathogens, such as viruses (e.g. iridescent and cytoplasmic polyhedrosis viruses), bacteria (e.g. *Bacillus thuringiensis* var. *israelensis* (= *B. thuringiensis* serotype H-14) and *B. sphaericus*), protozoans (e.g. *Bracheola* (= *Nosema*) *algerae* and *Vavraia culicis*) and fungi (e.g. species of *Coelomomyces*, *Lagenidium* and *Culicinomyces*) that cause larval mortality. There are also several parasitic nematodes that kill mosquito larvae, the best known of which is *Romanomermis culicivorax*, which at one time was commercially mass-produced, but because of non-viable sales is no longer obtainable commercially. All these parasites and pathogens appear harmless to humans, but have generally not been very successful biocontrol agents.

Bacillus thuringiensis var. *israelensis* (*Bti*) is undoubtedly the most useful pathogen, as it can be easily mass-produced, is toxicologically safe to humans and wildlife, and is more or less specific in killing mosquito larvae (but it can also kill simuliids: see Chapter 4). It is commonly formulated as a powder that after mixing with water is sprayed on breeding places, but because there is no multiplication of the bacteria there have to be repeated applications, as with most chemical larvicides. However, when formulated as slow-release granules or briquettes, insecticidal activity extends over many days, sometimes up to a month. When *Bti* is ingested mortality is caused by an endotoxin acting as a stomach poison. Genetic engineering is improving the larvicidal activity of the bacillus and in addition has transferred the genes responsible for production of the poisonous endotoxin to other bacteria.

Bacillus sphaericus can be formulated much as for *Bti* and kills mosquito larvae in a similar fashion, but differs in that in some situations it can recycle in larval habitats. This species is also more effective in organically polluted waters and is especially effective against *Culex* species. Both these *Bacillus* species are more like a microbial insecticide than a true biological (living) agent that recycles and maintains itself in the environment.

Resistance to both *Bti* and *B. sphaericus* has been observed in laboratory colonies of a few mosquito species, especially *Culex quinquefasciatus*. Field populations of this mosquito have been found to be resistant to *B. sphaericus* in Thailand, although they could be killed with *Bti*. There are, however, reports of field populations of *Cx. pipiens* in New York State being resistant to *Bti*. This serves as a warning that mosquitoes can become resistant to bacterial control agents.

Genetic control

Although genetic control methods are directed against the adults it is convenient to discuss this strategy here because it is really a form of biological control.

In the past genetic control involved releasing in the field sterile male mosquitoes that had been laboratory-reared. Sterilization was achieved by ionizing radiation, by crossing closely related species to produce infertile hybrid males, or more usually by introducing chemosterilants into larval breeding, which makes the emerging adults (both sexes) sterile. The aim was to introduce large numbers of healthy, but sterile, insectary-reared males into field populations that would compete with natural fertile males for female mates, so resulting in large numbers of infertile inseminations. Eggs layed by such females are sterile and fail to hatch, thus causing a reduction in, or even elimination of, the vector. In El Salvador the release in the 1970s of some 4.36 million chemosterilized male *Anopheles albimanus* (an important malaria vector that had developed resistance to most insecticides) over 4.5 months in an isolated coastal region of about 15 km^2 caused a more than 97% reduction in the biting population. However, because enormous numbers of mosquitoes had to be reared to obtain control over a very limited area, the later expanded trial failed. There is presently little interest in this technique for controlling mosquitoes.

Scientists skilled in genetic engineering are trying to create in the laboratory transgenic mosquitoes that are incapable of transmitting parasites, such as malaria plasmodia, and also viruses. Mechanisms must then be found to drive the genes through wild populations when the transgenic mosquitoes are released in field situations. However, none of these genetic approaches is simple, and they will be much more difficult to implement than conventional insecticidal methods. The logistics of this genetic approach and its ethics are currently being vigorously debated.

Physical control

Filling in, source reduction and drainage This is sometimes referred to as mechanical or environmental control. A simple form consists of filling in and thus completely eliminating breeding places. Larval habitats ranging in size from water-filled tree-holes to ponds and small marshes can be filled in with rubble, earth or sand. Filling in tree-holes can be difficult because many are at considerable heights and difficult to locate, or there can be too many for this method to be practical. Various container-habitats such as abandoned tin cans, metal drums, disused water-storage pots and old tyres can be removed: this approach is often referred to as *source reduction*. Mosquito breeding in water-storage pots that are in use can be reduced by covering up their openings, but this simple practice is often not popular and it soon becomes neglected. Introduction of a reliable piped water supply should help reduce people's dependence on water-storage containers and thereby reduce breeding of mosquitoes such as *Aedes aegypti*. However, in many areas having piped water people continue to store water in containers as insurance against an unreliable water supply. In the Indian subcontinent water tanks are commonly sited on rooftops and are important breeding places of the vector (*Anopheles stephensi*) of urban malaria. Fitting these with mosquito screening would prevent breeding, but such covers usually become torn or are removed.

Some mosquitoes, such as *Culex quinquefasciatus*, breed in damaged septic tanks and soakaway pits, but this is prevented if the tanks and pits are repaired so that egg-laying females cannot gain access. This mosquito also commonly breeds in pit latrines, but this can be stopped if small (2–3 mm) expanded **polystyrene beads** are tipped into latrines to form a floating layer about 1–2 cm thick. This suffocates larvae and pupae and also prevents mosquitoes laying their eggs on the water surface. In Zanzibar and India such beads substantially reduced *Cx. quinquefasciatus* breeding in pit latrines and soakage pits and contributed greatly to the prevention of a resurgence of bancroftian filariasis following mass drug treatmemnt.

Larval breeding places such as ponds, borrow pits, freshwater and salt-water marshes can be drained. An advantage of filling in, draining or removing larval habitats is that this can lead to permanent control, but this approach is not always feasible. It is impossible, for example, to fill in all the scattered, small and temporary collections of water such as pools, vehicle tracks and puddles which often appear during the rainy season. Larger and more permanent habitats such as swamps may prove too costly to drain. Moreover, the local people may, understandably, not want certain breeding places filled in if the water is needed for domestic purposes or the sites used as watering points for livestock. The feasibility of eradicating breeding places must be assessed individually in each area.

Environmental manipulation   If it is not feasible to eliminate mosquito breeding places it may be possible to alter them to make them unsuitable as larval habitats. For example, some mosquitoes breed in isolated pools and small marshy areas that form at the edges of ditches and streams having winding courses. Realigning these water courses to increase water flow and prevent the build-up of static pockets of water can greatly reduce mosquito breeding. The periodic opening of sluice gates has been practised in India and Malaysia to flush out larvae from small isolated pools of water. Other environmental modifications include the removal of over-hanging vegetation to reduce breeding by shade-loving mosquitoes; conversely planting vegetation along reservoirs and streams may eliminate sun-loving species. Intermittent flooding of ricefields to allow drying out every 3–5 days can substantially reduce populations of several important vectors. Removal of rooted or floating vegetation will prevent breeding of *Mansonia* species, because they require plants to obtain their oxygen requirements.

Instead of draining marshy areas they can be excavated to form areas of relatively deep permanent water with well-defined vertical banks. This process is called ***impoundment***. It completely alters the habitat, making it unsuitable for many mosquitoes, especially *Aedes* and *Psorophora* species, which lay their eggs on wet muddy edges of pools that are scattered over extensive marshy areas. Both small and large, freshwater and saltwater marshy areas can be converted into impounded waters. Sometimes such impounded waters are stocked with fish and ducks, which may also help reduce mosquito breeding.

There is the danger, however, that whenever a larval habitat has been modified to reduce breeding of certain mosquitoes the new conditions created may now support breeding of other mosquito species that were previously either absent or uncommon.

Chemical control

Most control directed against mosquitoes, except malaria vectors, consists of applying larvicides.

Oils Spraying mineral oils, such as diesel and kerosene (paraffin), to kill mosquito larvae has been used for over 100 years. The addition of detergents or 1–2.5% vegetable oils (e.g. castor oil, coconut oil) increases the spreading power of oils, allowing application rates to be reduced to 18–50 litres/ha. Although such oils are still used, mainly in tropical countries, elsewhere specially formulated commercial high-spreading oils which are environmentally more friendly are used. Dosage rates are commonly 9–27 litres/ha or less. Other larvicides include monomolecular films of ethoxylated isostearyl alcohol derived from plant oils (commercial names are Arosurf, Agnique), which interfere with the properties of the air–water interface and cause larvae, pupae and even emerging or ovipositing adults to drown.

Oils have to be sprayed on breeding places about every 7–10 days in most tropical countries to ensure that larvae hatching from eggs are killed before they pupate and give rise to adults. Less frequent applications are made in cooler temperate areas because the aquatic life cycle is much slower.

Insecticides With the availability of insecticides such as DDT in the mid 1940s, oiling was largely replaced with spraying larval habitats with these more modern chemicals. However, because of their persistence in the environment and accumulation in food chains DDT and other organochlorine insecticides should not now be used as larvicides. Less persistent and biodegradable insecticides should be used.

Recommended chemicals for larviciding include malathion, pirimiphos-methyl, chlorpyrifos, fenthion and temephos. Pyrethroids, such as permethrin and deltamethrin, can also be used as larvicides but because they tend to kill greater numbers of other aquatic insects, crustaceans and even fish they should be used with caution and only in special circumstances. In organically polluted waters insecticides are less effective and so either higher dosage rates must be used or the more effective organophosphates such as fenthion or chlorpyrifos applied. Chlorpyrifos is more toxic to mosquito larvae than many other insecticides, but also causes higher mortalities of fish and other aquatic organisms, so needs to be used with caution.

All the above insecticides usually have to be *sprayed* on breeding places in tropical areas every 10–14 days, and more frequently on highly polluted waters.

Temephos has very low mammalian toxicity, and briquettes, 1% sand granules or microencapsulated formulations which slowly release the insecticide over days or even weeks can be placed in containers holding potable water to control *Aedes aegypti*. However, people have sometimes refused to have their water pots treated with temephos, because of the unpleasant odour or else because they consider any insecticide in drinking water as environmental contamination. This attitude is likely to spread. In addition there are suggestions that temephos could be toxigenic and mutagenic.

Mansonia larvae can be killed by spraying herbicides to destroy the aquatic vegetation on which they rely to obtain their oxygen.

Larvicides are usually applied as emulsions or oil solutions, but granules (0.25–0.6 mm), pellets (0.6–2 mm) or gelatine capsules, often containing pyrethroids, are better for penetrating dense growths of aquatic vegetation. Insecticides formulated as *slow-release* granules or pellets can be scattered over marshy areas when they are relatively dry, and then when they become flooded larvae hatching from drought-resistant aedine eggs, such as those of *Aedes* and *Psorophora* species, are killed as the granules release their toxicants into the water. Larvicides are usually delivered from knapsack-type sprayers carried on the backs of operators, but they are sometimes dispersed from vehicle-mounted spraying machines. Large or inaccessible areas may require aerial spraying from helicopters or small fixed-wing aircraft.

Insect growth regulators (IGRs)

These are compounds, such as methoprene and pyriproxyfen, that arrest larval development of insects, or compounds, like diflubenzuron, which inhibit chitin formation in the immature stages. When used as larvicides these chemicals have the benefit of being environmentally friendly, because they are more or less specific in killing mosquitoes and possess extremely low toxicity to humans. Because of this methoprene has sometimes been used in water intended for drinking, but it is probably better to avoid this practice. Nevertheless, it is the most commonly used IGR for mosquito control, and can be formulated as liquids, granules or briquettes which can give control for more than 100 days. However, the relatively high cost of IGRs may limit their use in poorer countries.

Although resistance of mosquitoes to IGRs has not been reported, resistance to them has been documented in agricultural pests and in *Lucilia cuprina*, the sheep blowfly, which is of veterinary importance.

Integrated control
It has become fashionable to advocate *integrated* control, which usually means combining biological and insecticidal methods: for example, the introduction of predaceous fish to breeding places which are also sprayed with insecticides that have minimum effect on the fish. However, it is better to regard integrated control as any approach that takes into consideration more than one method, whether these are directed at only the larvae or the adults, or both.

1.5.2   Control directed at adults
Personal protection
Much can be done to reduce being bitten by mosquitoes. Houses, hospitals and other buildings can have windows and doors covered with mosquito screening, made of either strong plastic or non-corrosive metal. It is essential that screening is kept in good repair. Screens of 6–8 mesh (i.e. 6–8 holes/cm) will exclude most mosquitoes. Finer-mesh screening will keep out smaller biting flies, some of which may be vectors, but will appreciably reduce ventilation and light. If houses are unscreened, or if screening is defective, *mosquito nets* of 9–10 mesh can be used to protect against night-biting mosquitoes. The size of the holes depends on the thickness of the thread, as well as their number; generally their size is 1.2–1.5 mm. Nets should be tucked in under mattresses or bedding, never allowed to drape loosely over beds. Torn nets are useless unless they have been impregnated with pyrethroid insecticides such as permethrin (see p. 48). Nets should be placed over beds before sunset. The main disadvantage of nets is that they can reduce ventilation.

Small spray-guns (e.g. flit-guns) filled with pyrethrum or permethrin dissolved in kerosene, or alternatively pressurized aerosol canisters containing pyrethroids, are commonly used to spray bedrooms early in the evenings. Mosquito coils impregnated with pyrethroid insecticides, especially fast-acting ones like bioallethrin, which when ignited smoulder for 6–10 hours to produce an insecticidal smoke, are commonly used in tropical countries. A more sophisticated method is to place small insecticide-impregnated tablets (called vaporizing or fumigant mats) on a mains-operated electric mini-heater.

Suitable insect repellents are DEET, Autan (in North America known as 'Cutter Advanced') and Bayrepel. The latter two are based on piperidines, and are about as effective as DEET, but unlike DEET they do not attack plastics. Under optimal conditions these repellents can provide protection for 6–10 hours, the duration depending on the quantity of active ingredient. Citronella oil and lemon eucalyptus oil can give protection against mosquitoes, but only for about an hour. A new botanical repellent known as PMD (para-menthane 3,8-diol), derived from lemon eucalyptus, has proved about as good as DEET.

Repellents are applied to the hands, arms, neck and face (taking care to avoid the eyes), and the ankles and legs, irrespective of whether socks or long trousers are worn. Sweating and rubbing usually reduce the period of effectiveness of repellents. Commercially or locally made repellent soaps incorporating citronella oil or a mixture of DEET and permethrin can be effective. Repellent- or insecticide-impregnated (e.g. permethrin or allethrin) clothing, such as wide-mesh jackets and hoods (as used by military personnel) give longer protection than repellents applied to the skin. If treated clothing is kept in plastic bags when not in use it should remain effective for many months before re-impregnation is needed.

Aerosols, mists and fogs

Motorized knapsack mist-blowers, shoulder-carried thermal foggers or vehicle-mounted machines that generate insecticidal aerosols (< 50 μm) or mists (51–100 μm) can be used to kill outdoor-resting (exophilic) adult mosquitoes. Fogs are produced when very fine aerosol droplets (5–15 μm) are so numerous that they substantially reduce visibility. Indoor-resting (endophilic) adults are also occasionally killed by mist-blowers or thermal foggers. Several insecticides can be used, including malathion, fenitrothion, pirimiphos-methyl and the pyrethroids. Although such applications can be spectacular there is very little residual effect, and areas cleared of adult mosquitoes are rapidly invaded by newly emerged adults and mosquitoes flying in from outside the treated area. Repeated applications are needed to sustain control.

Applications of aerosols and mists are best made in calm weather, and usually in the evenings or early mornings when there are fewer thermals rising from the ground and less turbulence. Aerial applications from helicopters or fixed-wing aircraft usually give better coverage and more effective control than ground-based operations.

Ultra-low-volume applications Ultra-low-volume (ULV) techniques apply the minimum of concentrated insecticides, often just 225–500 ml/ha, as against 5–25 litres/ha with conventional spraying. This allows trucks or aircraft to spray much larger areas with a tank of insecticide before the tank needs refilling. Insecticides commonly used include malathion, pirimiphos-methyl and pyrethroids. With aerial applications *droplet size* of the insecticide is bigger (150–200 μm) than that used in ground-based applications (50–100 μm) because they decrease in size, due to evaporation, as they fall to the ground. Generally the size of droplets hitting mosquitoes should be 15–25 μm. In addition to rapidly reducing outdoor resting and biting mosquitoes, ULV spraying is used in potential or actual epidemic situations to prevent or control disease outbreaks. In emergency situations aerial spraying gives fast and effective vector control, and has been used to stop transmission of haemorrhagic dengue, Japanese encephalitis and in North America various encephalitis viruses.

Residual house-spraying

Some mosquitoes, such as many malaria vectors and *Culex quinquefasciatus*, rest in houses before and/or after blood-feeding. Their populations can be reduced by insecticidal spraying of houses, but as this approach is mainly used in malaria control operations it is described in Chapter 2.

Further reading

Anon. (1995) Vector control without chemicals: has it a future? A symposium. *Journal of the American Mosquito Control Association*, **11**, 247–93.

Bock, G. R. and Cardew, G. (eds.) (1996) *Olfaction in Mosquito–Host Interactions*. Chichester: Wiley.

Bowen, M. F. (1991) The sensory physiology of host-seeking behavior in mosquitoes. *Annual Review of Entomology*, **36**, 139–58.

Carrol, S. P. and Loye, J. (2006) PMD, a registered botanical mosquito repellent with DEET-like efficacy. *Journal of the American Mosquito Control Association*, **22**, 507–14.

Clark, G. G. (coordinator) (1994) Prevention of tropical diseases: status of new and emerging vector control strategies. Proceedings of a symposium on vector control. *American Journal of Tropical Medicine and Hygiene*, **50** (6) (suppl.), 1–159.

Clements, A. N. (1992) *The Biology of Mosquitoes. Volume 1: Development, Nutrition and Reproduction*. London: Chapman and Hall.

Clements, A. N. (1999) *The Biology of Mosquitoes. Volume 2: Sensory Reception and Behaviour*. Wallingford: CABI.

Curtis, C. F. (ed.) (1989) *Appropriate Technology in Vector Control*. Boca Raton, FL: CRC Press.

de Barjac, H. and Sutherland, D. J. (eds.) (1990) *Bacterial Control of Mosquitoes and Black Flies: Biochemistry, Genetics and Applications of Bacillus thuringiensis and Bacillus sphaericus*. New Brunswick, NJ: Rutgers University Press.

Foster, W. A. and Walker, E. D. (2002) Mosquitoes (Culicidae). In *Medical and Veterinary Entomology*, ed. G. Mullen and L. Durden. Amsterdam: Academic Press, pp. 203–62.

Harbach, R. and Sandlant, G. (1997) *CABIKEY: Mosquito Genera of the World*. CD-ROM for Windows 3x or Windows 95. Wallingford: CABI.

Horsfall, W. R. (1972) *Mosquitoes: Their Bionomics and Relation to Disease*. New York, NY: Hafner.

Lacey, L. A. and Lacey, C. M. (1990) The medical importance of riceland mosquitoes and their control using alternatives to chemical insecticides. *Journal of the American Mosquito Control Association*, **6** (suppl. 2), 1–93.

Laird, M. (1988) *The Natural History of Larval Mosquito Habitats*. London: Academic Press.

Laird, M. and Miles, J. W. (eds.) (1983) *Integrated Mosquito Control Methodologies. Volume 1: Experience and Components from Conventional Chemical Control*. London: Academic Press.

Laird, M. and Miles J. W. (eds.) (1985) *Integrated Mosquito Control Methodologies. Volume 2: Biocontrol and Other Innovative Components, and Future Directions*. London: Academic Press.

Pates, H. and Curtis, C. (2005) Mosquito behavior and vector control. *Annual Review of Entomology*, **50**, 53–70.

Schaffner, E., Angel, G., Geoffroy, B., Hervy, J.-P., Rhaiem, A. and Brunhes, J. (2000) *The Mosquitoes of Europe: an Identification and Training Programme*. CD-ROM. Windows 9x, NT, ME, 2000. Montpellier: EID. (In French and English.)

Service, M. W. (1989) Rice, a challenge to health. *Parasitology Today*, **5**, 162–5.

Service, M. W. (1993) Mosquitoes (Culicidae). In *Medical Insects and Arachnids*, ed. R. P. Lane and R. W. Crosskey. London: Chapman and Hall, pp. 120–40.

Service, M. W. (1993) *Mosquito Ecology: Field Sampling Methods*, 2nd edn. London: Chapman and Hall.

Spielman, A. and d'Antonio, M. (2001) *Mosquito: a Natural History of Our Most Persistent and Deadly Foe*. London: Faber and Faber.

Walter Reed Biosystematics Unit (2007, continually updated) Vector identification resources. www.wrbu.org.

World Health Organization (1990) *Equipment for Vector Control*, 3rd edn. Geneva: World Health Organization.

World Health Organization (1992) Vector resistance to pesticides. Fifteenth report of the WHO Expert Committee on Vector Biology and Control. *World Health Organization Technical Report Series*, **818**, 1–71.

World Health Organization (1996) *Operational Manual on the Application of Insecticides for Control of the Mosquito Vectors of Malaria and Other Diseases*. WHO/CTD/VBC/96.1000. Geneva: World Health Organization.

World Health Organization (1997) *Vector Control: Methods for Use by Individuals and Communities*, prepared by J. A. Rozendaal. Geneva: World Health Organization.

See also references at the ends of Chapters 2 and 3.

2

Anopheline mosquitoes (Anophelinae)

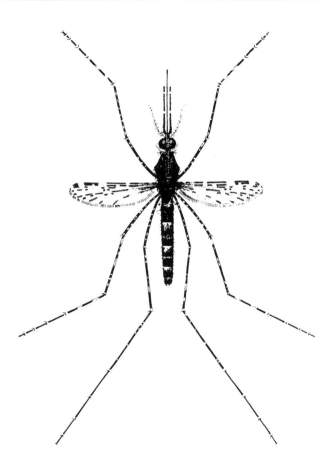

The subfamily Anophelinae contains three genera, but as explained in Chapter 1 only the genus *Anopheles* is of medical importance. *Anopheles* mosquitoes have a worldwide distribution, occurring in both tropical and temperate regions. There are about 484 species. The most important disease carried by *Anopheles* mosquitoes is malaria. Some *Anopheles* species are also vectors of filariasis, especially that caused by *Wuchereria bancrofti*, but some also transmit *Brugia malayi* and *Brugia timori*. A few species transmit arboviruses that are of minor medical importance.

2.1   External morphology

The main features distinguishing adults of the Anophelinae from other mosquitoes have been given in Chapter 1, but are briefly summarized here.

Eggs are laid singly and have air-filled *floats* (Fig. 1.8) that help them float on the water surface.

Larvae do not have a siphon and consequently lie *parallel* to the water surface (Fig. 1.13). Dorsally a *tergal plate* and paired *palmate hairs* are present on most abdominal segments (Fig. 1.9).

Pupal abdominal segments have numerous short setae, and segments 2–7 or 3–7 have in addition short peg-like spines (Fig. 1.11) which are absent in culicines.

Most, but not all, *Anopheles* have *spotted wings*, that is the dark and pale scales are arranged in small blocks or areas on the veins (Fig. 1.12, Plate 1). The number, length and arrangement of these dark and pale areas differ considerably in different species and provide useful characters for species identification. Unlike Culicinae the dorsal and ventral surfaces of the abdomen are almost, or entirely, devoid of appressed scales. In both sexes the *palps* are about as long as the proboscis, and in males, but not females, they are enlarged (that is clubbed) apically (Fig. 1.13). See page 18 for minor differences distinguishing anophelines from culicines.

2.2   Life cycle

After mating and blood-feeding *Anopheles* lay some 50–200 small brown or blackish boat-shaped eggs (Fig. 1.8) on the water surface. *Anopheles* eggs cannot withstand desiccation and in tropical countries hatch within 2–3 days, but in colder temperate climates hatching may not occur until after about 2–3 weeks, the duration depending on temperature.

As in all mosquitoes there are four larval instars. *Anopheles* larvae are filter-feeders and unless disturbed remain at the water surface, feeding on bacteria, yeasts, protozoans and other micro-organisms, and also breathing in air through their spiracles. When feeding, larvae rotate their heads through 180° so that the ventrally positioned mouthbrushes can sweep the underside of the water surface. Larvae are easily disturbed by shadows or vibrations and respond by swimming quickly to the bottom of the water. They resurface some seconds or minutes later.

Anopheles larvae occur in many different types of more or less permanent habitats, ranging from freshwater and saltwater marshes, mangrove swamps, grassy ditches, ricefields, wells, edges of streams and rivers to ponds and borrow pits. They are also found in small and often temporary breeding places such as puddles, hoofprints, discarded tin cans and sometimes water-storage pots. A few species occur in water-filled treeholes. In the Neotropical region (Central and South America and the West Indies) a few *Anopheles* breed in water that collects in the leaf axils of epiphytic plants such as **bromeliads** growing on tree branches. Bromeliads somewhat resemble pineapple plants. Some *Anopheles* prefer habitats with aquatic vegetation while others favour habitats without vegetation; some species like exposed sunlit waters whereas others prefer more shaded breeding places. In general *Anopheles* are found in clean and unpolluted waters, being usually absent from habitats containing rotting plants or faeces.

In tropical countries the larval period frequently lasts only about 7 days, but in cooler climates the larval period may be 2–4 weeks. In temperate areas some *Anopheles* overwinter as larvae and consequently may live many months.

The comma-shaped pupae normally remain floating at the water surface but when disturbed they swim vigorously down to the bottom with characteristic jerky movements. The pupal period lasts 2–3 days in tropical countries but sometimes as long as 1–2 weeks in cooler climates, after which the adult mosquito emerges.

2.2.1 Adult biology and behaviour

Most *Anopheles* are **crepuscular** or **nocturnal** in their activities. Thus bloodfeeding and oviposition normally occur in the evenings, at night or in the early mornings around sunrise. Some species such as *An. albimanus*, a malaria vector in Central and South America, bite people mainly outdoors (**exophagic**) from about sunset to 21:00 hours. In contrast, in Africa species of the *An. gambiae* complex, which contains probably the world's most efficient malaria vectors, bite mainly after 23:00 hours and mostly indoors (**endophagic**). As already discussed in Chapter 1, the times of biting, and whether adult mosquitoes are exophagic or endophagic, may be epidemiologically important.

Both before and after blood-feeding some species rest in houses (**endophilic**), whereas others rest outdoors (**exophilic**) in a variety of shelters, such as amongst vegetation, in rodent burrows, in cracks and crevices in trees, under bridges, in termite mounds, in caves and rock fissures, and in cracks in the ground. Most *Anopheles* species are not exclusively exophagic or endophagic, exophilic or endophilic, but exhibit a mixture of these extremes of behaviour. Similarly, few *Anopheles* feed exclusively on either humans or non-humans, most feeding on both people and animals, but the degree of anthropophagism and zoophagism varies according to species. For example, *An. culicifacies*, an important Indian malaria vector, commonly feeds on cattle as well as humans, whereas in Africa *An. gambiae* sensu stricto (a species of the *gambiae* complex) feeds more rarely on cattle and thus maintains a stronger mosquito–man contact. This is one of the reasons why *An. gambiae* is a more efficient malaria vector than *An. culicifacies*.

2.3   Medical importance

2.3.1   Biting nuisance

Although *Anopheles* mosquitoes may not be disease vectors in an area they may nevertheless be a biting nuisance. Usually, however, it is culicine mosquitoes, especially *Aedes* and *Psorophora* species, that cause biting problems.

2.3.2   Malaria

Only mosquitoes of the genus *Anopheles* transmit the four parasites (*Plasmodium falciparum*, *P. vivax*, *P. malariae*, *P. ovale*) that cause human malaria. Because the sexual cycle of the malaria parasite occurs in the vector, it is conventional to call the mosquito the **definitive host**, and humans the intermediate host.

Male and female malaria gametocytes ingested by female mosquitoes during blood-feeding are passed to the mosquito's stomach, where they undergo cyclical development that includes a sexual cycle termed sporogony. Only **gametocytes** survive in the mosquito's stomach; all other blood forms of the malaria parasites (the asexual forms) are destroyed. Male gametocytes (microgametocytes) extrude flagella which are the male gametes (microgametes), and the process is called **exflagellation**. The microgametes break free and fertilize the female gametes (macrogametes) which have formed from the macrogametocytes. As a result of fertilization a **zygote** is formed, which elongates to become an ookinete. This penetrates the wall of the mosquito's stomach and reaches its outer membrane where it becomes spherical and develops into an **oocyst**, which can be seen on the stomach walls of vectors about 4–5 days after an infective blood-meal. The nucleus of the oocyst divides repeatedly to produce numerous spindle-shaped **sporozoites**. After about 8 days the oocyst is fully grown (about 40–80 μm) and it ruptures to release thousands of sporozoites into the haemocoel of the mosquito. The sporozoites (10–15 μm) are carried in the insect's haemolymph to all parts of the body but most penetrate the **salivary glands**, where they are usually found after 9–14 days. However, the time required for this cyclical development (**extrinsic cycle**) depends on both temperature and *Plasmodium* species. At 30 °C sporogony in

P. falciparum takes 9 days, at 25 °C 10 days, and at 20 °C 23 days, while below 17 °C it cannot be completed. With *Plasmodium vivax* sporogony develops faster: it is completed in 9 days at 25 °C and in 16 days at 20 °C.

The mosquito is now *infective* and sporozoites are inoculated into people the next time the mosquito bites. A single oocyst produces 1000 or more sporozoites, and it has been estimated that in heavy infections there may be as many as 60 000–70 000 sporozoites in the vector's salivary glands. However, the number may be much smaller than this, and very few (sometimes just 5–10) are actually injected into a person during feeding.

The *sporozoite rate*, that is the percentage of female vectors with sporozoites in their salivary glands, varies considerably, not only from species to species of mosquito but also according to locality and season. Sporozoite rates are often about 1–5% in species such as *An. gambiae* and *An. arabiensis* of the *An. gambiae* complex, but less than 1% in many other species such as *An. albimanus* and *An. culicifacies*. For practical purposes it can be said that once a vector becomes infective it remains so throughout its life.

2.3.3   Important malaria vectors

Although there are some 484 species of *Anopheles* only about 70 are malaria vectors, and of these probably only about 40 are important ones. Malaria vectors are often divided into primary and secondary vectors, but this can be misleading because a species may be considered a primary vector in some areas but only a secondary vector in others. Although presenting a list of important malaria vectors must be somewhat subjective I have nevertheless attempted to do this, and I provide notes on their principal larval habitats and biting behaviour. These notes are only a guide to their behaviour, which may vary in different parts of each species' geographical range. Several species occur as species complexes, which comprise virtually identical-looking species that can be separated only by their chromosomal banding patterns, by biochemical procedures or by molecular methods. However, species within a complex may differ in their behaviour, distribution and vector status. Some of the more important species complexes are referred to in this list of vectors, of which the most important and best known is the *An. gambiae* complex in Africa.

(1) Sub-Saharan Africa

Anopheles gambiae The most anthropophagic species and the most important malaria vector of the seven that comprise the *An. gambiae* complex. Larval habitats are sunlit pools, puddles, hoofprints, borrow pits and ricefields. Adults bite humans both indoors and outdoors, and also feed on domesticated animals. They rest mainly indoors, but also outdoors. Other malaria vectors in the *An. gambiae* complex are *An. arabiensis*, *An. melas*, *An. merus* and *An. bwambae* (see below). *Anopheles quadriannulatus* species

A and B are also in the complex, but as they bite cattle in preference to humans they are not malaria vectors.

Anopheles arabiensis Another important malaria vector. Larvae in same habitats as *An. gambiae*. Adults bite humans indoors and outdoors but also cattle; after feeding they rest either indoors or outdoors. This species tends to occur in drier areas than does *An. gambiae*, and is more likely to bite cattle and rest outdoors than *An. gambiae*.

Anopheles melas and *Anopheles merus* *Anopheles melas* breeds in coastal salt waters such as mangrove swamps in West Africa. *Anopheles merus* is a coastal saltwater species in East and southern Africa, but can also be found breeding in inland saltwater habitats. Adults of both species bite humans and rest indoors and outdoors; they are both regarded as secondary malaria vectors.

Anopheles bwambae A rare species restricted to breeding in the warm mineral springs in Semuliki National Park, Uganda. Not considered an important vector, although locally it can transmit malaria.

Anopheles funestus Larvae occur in more or less permanent waters, especially those with vegetation, such as marshes, edges of streams, rivers and ditches, and ricefields with mature plants providing shade. Prefers shaded habitats. Adults bite humans predominantly, but also domesticated animals; feeds indoors and also outdoors; after feeding adults rest mainly indoors.

(2) Europe, North Africa and the Middle East
Anopheles atroparvus One of 12 species in the *An. maculipennis* complex. Sunlit and exposed pools and ditches with either fresh or brackish water, also ricefields. Adults bite humans and domesticated animals and usually rest in stables, cowsheds and piggeries. Adults hibernate in these and other shelters during the winter, but periodically emerge to take blood-meals.

Anopheles labranchiae Another of the 12 species in the *An. maculipennis* complex. Brackish waters of coastal marshes, freshwater marshes, edges of grassy streams and ditches, and ricefields; prefers sunlight. Bites humans and domesticated animals indoors and outdoors; rests mainly in houses or animal shelters after feeding. Overwinters as hibernating adults.

Anopheles pharoensis Marshes, ponds, especially those with abundant grassy or floating vegetation, also ricefields. Adults bite humans and animals indoors or outdoors, and rest outdoors after feeding. Can be an important vector in Egypt.

Anopheles sacharovi Fresh or brackish waters of coastal or inland marshes, pools and ponds, especially those with vegetation. Prefers sunlit habitats. Bites humans and animals indoors or outdoors; usually rests in houses or animal shelters after feeding.

Anopheles sergentii Borrow pits, ricefields, ditches, seepage waters, slow-flowing streams, sunlit or partially shaded habitats. Adults bite humans or animals indoors or outdoors, resting in houses and caves after feeding.

Anopheles stephensi Possibly a species within a complex. Can be an important vector locally, especially in towns. Apart from being found in Egypt, Iraq, Iran and Saudi Arabia it is common in the Indian subcontinent, where it is commonly the main vector of urban malaria. Larvae breed in fresh, brackish or even polluted waters in man-made habitats such as water tanks, cisterns, wells, gutters, water-storage jars and containers. Adults bite humans indoors or outdoors, and rest mainly indoors afterwards. *Anopheles stephensi* has a very wide distribution extending from the Middle East across Pakistan and India, to Myanmar, Thailand and China.

Anopheles superpictus Flowing waters such as torrents of shallow water over rocky streams, pools in rivers, muddy hill streams. Vegetation may be present; prefers sunlight. Bites humans and animals indoors and outdoors, and after feeding rests mainly in houses and animal shelters, but also in caves.

(3) Indian subcontinent
Anopheles annularis Can be an important vector in India. Larvae in ponds, especially those with vegetation, swamps and ricefields. Adults bite humans and cattle outdoors and indoors, and rest mainly outdoors after feeding.

Anopheles culicifacies One of five species in the *An. culicifacies* complex. Most important vector in the Indian subcontinent. Larvae in a variety of clean and unpolluted habitats, irrigation ditches, pools, wells, borrow pits, edges of streams, marshes, ricefields and occasionally in brackish waters. Adults prefer domesticated animals but commonly bite humans indoors or outdoors; rests mainly indoors after feeding. The main malaria vector in much of the region.

Anopheles fluviatilis One of two species in the *An. fluviatilis* complex. Flowing waters such as hill streams, pools in riverbeds, irrigation ditches, seepages; prefers sunlight. Bites humans and domesticated animals; feeds and rests either indoors or outdoors.

Anopheles minimus One of three species in the *An. minimus* complex. Flowing waters such as foothill streams, springs, irrigation ditches, seepages, borrow pits and ricefields; prefers shaded areas. Feeds mainly on humans, but will bite domesticated animals; feeds and rests mainly indoors.

Anopheles stephensi See Europe, North Africa and the Middle East, above.

Anopheles sundaicus One of four species in the *An. sundaicus* complex. Salt or brackish waters including lagoons, marshes, pools and seepages, especially with putrefying algae and aquatic weeds. Mainly a coastal species but found in freshwater inland pools in Java and Sumatra. Prefers sunlit habitats. Bites humans and domesticated animals indoors and outdoors; rests mainly indoors after feeding.

Anopheles superpictus See Europe, North Africa and the Middle East, above.

(4) South-east Asia

Anopheles aconitus Ricefields, swamps, irrigation ditches, pools and streams with vegetation; prefers sunlit habitats. Adults feed indoors or outdoors on humans but also commonly on cattle; adults rest indoors or outdoors after feeding.

Anopheles anthropophagus Shaded pools and ponds, only rarely in ricefields. Adults bite humans indoors and rest mainly indoors after feeding.

Anopheles balabacensis Muddy and shaded forest pools, animal hoofprints and vehicle ruts, occasionally in deep wells. Adults bite humans and cattle and feed and rest outdoors. Morphologically and biologically very similar to *An. dirus*, but has a more restricted distribution – Sabah, Java, Borneo and certain Philippine islands.

Anopheles campestris Deep and usually shaded or partially shaded waters such as ditches, wells, and shaded parts of ricefields, but also sometimes in brackish waters. Bites humans and animals indoors and outdoors, and rests indoors or outdoors after feeding.

Anopheles culicifacies See Indian subcontinent, above.

Anopheles dirus One of at least seven species in the *An. dirus* complex. Shaded pools, hoofprints in or at the edges of forests. Adults bite humans and domesticated animals, mainly outdoors, and stay outdoors after feeding. Morphologically and biologically very similar to *An. balabacensis*, but has a more widespread distribution from western India and Bangladesh to South-east Asia.

Anopheles donaldi Shaded habitats such as tree-covered swamps, forest pools, often with vegetation, ricefields. Bites domesticated animals but also feeds on humans inside or outside houses; rests mainly outdoors.

Anopheles flavirostris Flowing waters such as foothill streams, springs, irrigation ditches, also borrow pits and ricefields. Prefers shaded areas of sunlit habitats. Feeds mainly on humans, but also on domesticated animals; feeds indoors or outdoors but rests mainly outdoors after feeding.

Anopheles fluviatilis See Indian subcontinent, above.

Anopheles letifer Often in acidic and stagnant pools, swamps and ponds, especially on coastal plains; prefers shade. Adults bite domesticated animals and humans mainly outdoors, and rest afterwards outdoors.

Anopheles leucosphyrus One of two species in the *An. leucosphyrus* complex. Clear seepage pools in forests. Adults bite humans inside and outside houses, but afterwards rest outdoors. This species is morphologically and biologically similar to *An. balabacensis* and *An. dirus*.

Anopheles maculatus Seepage waters, pools formed in streams, rockpools, edges of ponds, ditches and swamps with much vegetation; prefers sunlight. Bites humans and animals mainly outdoors and rests outdoors after feeding.

Anopheles minimus See Indian subcontinent, above.

Anopheles nigerrimus Larvae in deep ponds, ricefields, irrigation ditches and marshes having much vegetation; prefers sunlight. Adults bite humans and animals mainly outdoors, and rest mainly outdoors.

Anopheles sinensis Has been often confused with *An. anthropophagus*. Common in China, where in some localities it may be a more important vector than *An. anthropophagus*. Larvae in ricefields, marshes, ditches and grassy ponds. Adults bite cattle and humans, indoors or outdoors; rests indoors or outdoors.

Anopheles subpictus Muddy pools near houses, borrow pits, gutters, also brackish waters. Bites mainly animals, but also humans both indoors and outdoors; rests indoors or outdoors after feeding.

Anopheles sundaicus See Indian subcontinent, above.

(5) Mexico and Central America

Anopheles albimanus Fresh or brackish waters such as pools, puddles, marshes, ponds, ricefields and lagoons, especially those containing floating or grassy vegetation; prefers sunlit habitats. Adults feed on humans and domesticated animals both indoors and outdoors; after feeding adults rest mainly indoors.

Anopheles albitarsis Larvae usually in sunlit ponds, large pools and marshes with filamentous algae. Bites humans and domesticated animals almost indiscriminately. Feeds outdoors and also indoors, and usually rests outdoors after feeding.

Anopheles aquasalis Tidal saltwater marshes, lagoons, saltwater regions of rivers, estuaries, rarely in fresh water; sunlit or shaded habitats. Adults bite humans and domesticated animals indoors or outdoors, rests mainly outdoors.

Anopheles darlingi Freshwater marshes, lagoons, ricefields, swamps, lakes, ponds, pools, edges of streams, especially with vegetation. Mainly shaded larval habitats. Feeds mainly on humans indoors; stays indoors after feeding.

Anopheles pseudopunctipennis Pools, puddles, seepage waters and edges of streams, especially habitats with algae; prefers sunlight. Feeds almost indiscriminately indoors or outdoors on humans and domesticated animals; rests outdoors.

Anopheles punctimacula Small pools, swamps, grassy pools at edges of streams; prefers shade. Bites humans and domesticated animals both indoors and outdoors; rests indoors or outdoors after feeding.

(6) South America

Anopheles albitarsis See Mexico and Central America, above.

Anopheles aquasalis See Mexico and Central America, above.

Anopheles bellator Larvae found only in water collected in the leaf axils of bromeliads, which are epiphytes on trees in the Americas. Adults bite humans during the day in shaded forests, and also at night, and may enter houses to feed. Adults rest mainly outdoors. The species will also bite domesticated animals. It occurs in Trinidad, Venezuela, Surinam, Guyana and Brazil, where it can be a local malaria vector.

Anopheles cruzii Another malaria vector that also breeds in bromeliad axils. Adults bite humans indoors and outdoors, and rest indoors and outdoors. Mainly a malaria vector in coastal areas of Brazil.

Anopheles darlingi See Mexico and Central America, above.

Anopheles nuneztovari One of three species in the *An. nuneztovari* complex. Muddy waters of pools, vehicle tracks, hoofprints and small ponds, especially in and around towns; prefers sunlight. Feeds mainly on animals but in northern Colombia and western Venezuela adults bite humans indoors and outdoors, and rest outdoors after feeding.

Anopheles punctimacula See Mexico and Central America, above.

(7) Australasia

Anopheles farauti *Anopheles farauti* is one of seven very closely related species forming the *farauti* complex, which together form part of the 12 species in the *An. punctulatus* group. Larvae usually in semi-permanent waters such as swamps, ponds, lagoons and edges of slow-flowing streams, but also in puddles, hoofprints, pools and rarely in man-made containers; water may be fresh or brackish and in sunlight or shade. Adults bite animals but also humans indoors or outdoors, and rest mainly outdoors, but also indoors.

Anopheles koliensis A species in the *An. punctulatus* group. Larvae occur in marshy pools, irrigation ditches, pools at edges of forest streams, often in sunlit habitats. Adults bite humans and more occasionally animals, and after feeding rest mainly indoors, but they rest outdoors in some areas.

Anopheles punctulatus One of two very similar species in the *An. punctulatus* complex. Larvae occur in temporary and often muddy pools, puddles, hoofprints and ditches, usually in sunlight. Adults bite humans in preference to animals, and rest indoors or outdoors after feeding.

2.3.4 Filariasis

Certain *Anopheles* species transmit filarial worms of *Wuchereria bancrofti*, *Brugia malayi* and *Brugia timori*, all of which cause filariasis in humans. The role of culicine mosquitoes as vectors of the first two species is described in Chapter 3, and details of principal mosquito vectors are given in Table 3.1.

Wuchereria bancrofti is the most widespread filarial infection of humans in many subtropical and tropical countries in Africa, Asia, the South Pacific and the Americas. In many of these areas bancroftian filariasis is mainly an urban disease. In contrast *B. malayi* is more a rural disease and has a much more restricted distribution, occurring in Asian countries such as southern India, Malaysia, Vietnam, Indonesia, Thailand, Papua New Guinea and the Philippines. It is absent from Africa and the Americas.

Both bancroftian and brugian filariasis occur in two basic forms: *nocturnally periodic* and *nocturnally subperiodic*. Anophelines, together with certain culicines (*Culex quinquefasciatus*, various species of *Mansonia* and a few *Aedes* species) are the vectors of the nocturnally periodic form. Culicines are vectors of subperiodic forms (see Chapter 3).

In the nocturnally periodic form of these two parasites most of the microfilariae during the day are in the blood vessels supplying the lungs. At night, especially during the middle part, microfilariae migrate to the peripheral blood system and lymph vessels. Because of this marked 24-hour periodicity microfilariae are ingested mainly by night-biting mosquitoes such as *Anopheles*.

More than 25 anopheline species are known vectors of bancroftian filariasis. The species involved differ according to the area, and many are also the principal vectors of malaria. For example, vectors of nocturnally periodic *W. bancrofti* include *An. albimanus* (tropical Americas), *An. arabiensis, An. funestus, An. gambiae* (sub-Saharan Africa), *An. anthropophagus, An. balabacensis, An. dirus, An. flavirostris, An. letifer, An. leucosphyrus, An. maculatus, An. sinensis* and *An. subpictus* (India and South-east Asia), and *An. koliensis* and *An. punctulatus* (Papua New Guinea). There are no known animal reservoir hosts of the nocturnally periodic form of filariasis.

Nocturnally periodic *B. malayi* is widespread in Asia, where it is primarily a rural disease. Transmission is by culicine mosquitoes and at least 10 anopheline species, including *An. anthropophagus, An. barbirostris, An. campestris, An. donaldi* and *An. sinensis*. There are no important animal reservoir hosts, although it is possible that some exist.

Brugia timori is known only from the small Indonesian islands of Alor, Timor and Flores, and from lowland areas of small neighbouring islands east of Java. Its microfilariae are nocturnally periodic and are transmitted by *An. barbirostris*, and possibly by other anopheline species. There are no known animal reservoir hosts.

Filarial development in mosquitoes

Development of all three filarial species and their mode of transmission from mosquito to humans are the same for all vectors. Basically, the life cycle in the mosquito is as follows. *Microfilariae* ingested with a blood-meal pass into the mosquito's stomach (in some vectors such as *Anopheles* many may be destroyed during their passage through the oesophagus).

Within 15–30 minutes they exsheath, penetrate the stomach wall and pass into the *haemocoel*, from where they migrate to the mosquito's *thoracic muscles*. In the thorax the small larvae become more or less inactive, grow shorter but considerably fatter and develop, after 2 days, into 'sausage-shaped' forms. They undergo two moults and the resultant *third-stage* larvae become active, leave the muscles and migrate through the head and down the fleshy labium of the proboscis. This is the infective stage and is formed some 10 days or more after the microfilariae have been ingested with a blood-meal.

When the mosquito takes further blood-meals, several infective (third-stage) larvae (1.2–1.6 mm long) emerge from the labium and crawl onto the surface of the host's skin. However, many of these die and only a few find a skin abrasion, sometimes the small lesion caused by the mosquito's bite, and enter the skin and pass to the host's lymphatic system. The salivary glands are not involved in the transmission of filariasis, and there is no multiplication or sexual cycle of the parasites in the mosquito.

Infection rates of infective larvae in anopheline vectors vary according to the mosquito species and local conditions, but they are often about 0.1–5% for *W. bancrofti* and about 0.1–3% for *B. malayi*.

There are no animal reservoir hosts of *W. bancrofti*, but the nocturnally subperiodic form of *B. malayi*, transmitted by *Mansonia* mosquitoes, is a zoonosis (see Chapter 3).

The presence of filarial worms in the mosquito's thoracic muscles, or infective worms in the proboscis, does not necessarily implicate mosquitoes as vectors of bancroftian or brugian filariasis. This is because there are several other mosquito-transmitted filariae. For example, various *Setaria* species infect cattle, *Dirofilaria repens* and *Dirofilaria immitis* infect dogs and humans, and other species of *Brugia*, such as *B. patei* in Africa and *B. pahangi* in Asia, infect animals not people. Careful examination is therefore essential to identify whether filarial parasites found in mosquitoes are those of *W. bancrofti* or *B. malayi*.

Table 3.1 (p. 74) summarizes the distribution and vectors of filariasis.

2.3.5 Arboviruses

The word arbovirus is derived from the term '*ar*thropod-*bo*rne *virus*'. An arbovirus infection in either human or non-human hosts produces *viraemia* and the virus is ingested by blood-sucking insects such as mosquitoes when they take blood-meals. Within the vector the virus undergoes multiplication and/or cyclical development before being transmitted by the infected arthropod during refeeding. An arbovirus therefore undergoes obligatory development in an arthropod host. Yellow fever and dengue are typical arboviruses transmitted by *Aedes* mosquitoes. In contrast the virus causing poliomyelitis is not an arbovirus, for although it can be transmitted by house-flies this is purely mechanical transmission, that is the virus does not undergo any multiplication and/or development in an arthropod. Time taken for an infected mosquito to become *infective*, that is having virus in the salivary glands, is called the ***extrinsic incubation period***, and varies according to temperature and the species of both arbovirus and mosquito. Over 500 arboviruses have been catalogued, but little is known about the epidemiology of many of them, and not all may be real arboviruses. About 150 arboviruses infecting humans are transmitted by mosquitoes, mostly by culicines, in particular by *Aedes* and *Culex* species. Other arboviruses are spread by other arthropods, especially ticks.

In 1959 a new painful, but non-fatal, arboviral disease called o'nyong nyong (a local African word meaning 'joint-breaker') was isolated from a patient in Uganda. In 1960–61 a major epidemic of this *Alphavirus* was identified and the disease spread to Kenya and other countries of East and Central Africa. The virus was found to be transmitted by the *An. gambiae* complex and *An. funestus*. This was the first time an *Anopheles* mosquito was incriminated as a vector of an arbovirus. Since then about 20 other arboviruses infecting humans have been found to be transmitted by *Anopheles*, none being epidemiologically very important.

2.4   Control
The principal methods of mosquito control are described in Chapter 1. Here only those methods specifically directed against *Anopheles*, mainly because of their role as malaria vectors, are considered.

2.4.1   Larval control
Until the availability of DDT and other organochlorines in the mid 1940s, when spraying houses with insecticides became the main strategy in malaria control programmes, larviciding with petroleum oils and Paris Green was widely practised. When properly used, larvicides reduced malaria in localized areas of economic or social importance such as principal towns, coffee and tea estates, rubber plantations and mining camps. However, because spraying had to be repeated about every 7–14 days it was logistically impossible to control malaria over large rural areas. This only became possible with the introduction of house-spraying. Nevertheless, when larval habitats are readily identified, larviciding with insecticides or applying insect growth regulators such as pyriproxyfen and the microbial insecticide *Bacillus thuringiensis* var. *israelensis* may be appropriate as a part of malaria control campaigns.

Predatory fish, mainly *Gambusia* species, continue to be used to reduce larval populations in some areas. However, they have often not been very effective in reducing malaria transmission, but they are more likely to reduce malaria in arid areas in countries such as Iran, Afghanistan, Somalia and Ethiopia, where larval habitats are discrete and permanent. In Kanarka State, India, where the malaria vector *An. culicifacies* species A breeds in wells and ponds, introducing fish (*Poecilia reticulata*) in these habitats has significantly reduced malaria, while in Goa *An. stephensi* larvae in wells have been controlled with fish.

Draining swamps and marshes has sometimes reduced *Anopheles* breeding. However, draining or filling in breeding places is impractical in many situations, for example with species breeding in temporary habitats such as scattered pools and puddles, or those breeding in forest pools and swamps.

2.4.2 Adult control
In most malaria control campaigns, control is now focused on the adults.

Residual house-spraying
This widely practised method consists of spraying water-dispersible (wettable) powders of residual insecticides to the interior surfaces of walls, ceilings and roofs of houses. DDT is an effective insecticide, if vectors have not developed resistance to it. However, because DDT accumulates in mammalian tissues and residues have been found in human breast milk many countries banned it. However, presence of residues (which may be minute) does not necessarily mean that human health is affected. In 2006 both the World Health Organization and the United States Agency for International Development (USAID) endorsed the use of DDT for spraying houses to control malaria. DDT, generally in the form of a water-dispersible powder at the rate of 2 g/m^2, is now being used again in several African countries, such as Cameroon, Eritrea, Ethiopia, Mozambique, Namibia, South Africa, Swaziland, Uganda, Zambia and Zimbabwe; Asian countries like India, Myanmar and Thailand; and in the Americas in Venezuela. Even in highly endemic areas houses need be sprayed only at six-monthly intervals, and where malaria transmission is very seasonal, such as just during the monsoon season, a single spraying before the rains may be sufficient to give good control.

If mosquitoes are resistant to DDT then organophosphates such as malathion, fenitrothion or pirimiphos-methyl or the carbamate propoxur can be used, usually at rates of $1.5–2 \text{ g/m}^2$. Bendiocarb, another carbamate, has a lower dosage rate of just $0.1–0.4 \text{ g/m}^2$. Alternatively pyrethroids such as lambda-cyhalothrin or deltamethrin, both at $0.02–0.03 \text{ g/m}^2$, or permethrin at 0.5 g/m^2 can be used. However, these alternative insecticides are less persistent, and spraying may have to be repeated every 3–4 months. They are also more expensive than DDT, but sometimes the lower dosage rates compensate for this. Most are also more toxic to people than DDT, and consequently stricter safety measures must be introduced during control programmes.

House-spraying is often popular because it also kills bedbugs, houseflies and cockroaches. However, the effectiveness of spraying houses in malaria control programmes depends on the mosquitoes resting indoors, and many important vectors, especially those in South-east Asia and the tropical Americas, feed and/or rest out of doors. House-spraying may promote the selection of exophilic populations of a species, that is populations that before spraying rested in houses now rest out of doors. Similarly, spraying may reduce populations of one species of a complex that is primarily endophilic, but by so doing allow an increase in numbers of another species in the complex that is exophilic. Such population changes have been recorded in the *An. gambiae* complex in Zimbabwe, *An. nuneztovari* in Venezuela, *An. minimus* in Thailand and in a few other species.

Insecticide-treated nets (ITNs)

Many of the bed-nets bought by poor communities are cheap, badly made and soon tear, and torn nets are useless in protecting people against mosquito bites. Also, mosquitoes will readily bite through a net if any part of the body is pressed up against it. However, nets that are impregnated with pyrethroid insecticides such as permethrin (200 mg/m^2), deltamethrin (25 mg/m^2), alpha-cypermethrin (20 mg/m^2), lambda-cyhalothrin (10 mg/m^2), cyfluthrin (50 mg/m^2) and etofenprox (200 mg/m^2), all of which repel and kill mosquitoes, will still protect people against bites even if nets are torn or people sleep up against them. Nets can be impregnated by simply dipping them in insecticide contained in a plastic bowl or dustbin. Such impregnated nets can often remain effective for about a year before they need re-dipping in insecticide. Washing nets tends to make them less effective, although nets treated with some pyrethroids, such as alpha-cypermethrin, can remain effective after at least five washes.

Long-lasting insecticidal nets (LLINs) can be made by incorporating permethrin into polyethylene during manufacture and before it is made into fibres ('Olyset nets').

Alternatively during manufacture deltamethrin can be glued onto the fibres with a resin to give long-lasting nets ('Permanets'). Sachets of a pyrethroid insecticide and a resin 'adhesive' provide a means by which home treatment can make any net a long-lasting one.

In China 38 million Permanets have been distributed to the public during measles vaccination campaigns. Long-lasting nets are expected to remain effective until the end of their physical 'life', when they have to be replaced by new ones.

It appears that if used on a large scale impregnated nets can have a mass-killing effect on vector populations which can benefit the whole community, even those without nets. Many regard such nets as the main hope for malaria control; however, resistance of malaria vectors (e.g. *An. gambiae* s.l. and *An. funestus*) to some pyrethroids has already been documented and will likely increase. Because of this, especially in sub-Saharan Africa, experiments with DEET-treated bed-nets have been undertaken. But such nets remain effective for only up to six weeks.

Nevertheless the future seems to be in the use of long-lasting insecticidal nets (LLINs) that would greatly reduce, if not make obsolete, the need to re-dip them. International and other agencies planned to have distributed 38 399 000 such nets to sub-Saharan African countries by 2007.

Nets will also give variable protection against mosquito vectors of filariasis and phlebotomine sand-fly vectors of leishmaniasis (Chapter 5), but will not be effective against vectors that bite out of doors or in the early evening before people have gone to bed.

2.4.3   Malaria control and malaria eradication

In 1955 the eighth World Health Assembly stated that worldwide malaria eradication, except in Africa south of the Sahara, was technically feasible. However, a sense of urgency in achieving this aim was recognized because insecticide resistance in *Anopheles* had been reported in 1950. In 1968 the 22nd World Health Assembly realized it had been over-optimistic and declared that global malaria eradication was not at present possible, although it remained the ultimate goal, and that for the time being malaria control should be the aim. This basically remains the situation today.

The difference between malaria eradication and malaria control is that eradication means the total cessation of transmission and elimination of the reservoir of infection in people so that at the end of the antimalaria campaign there is no resumption of transmission. Malaria control means reducing malaria transmission to an acceptable rate, that is to a level that no longer constitutes a major public health problem. For this control measures have to be maintained indefinitely; if they are relaxed malaria prevalence will rise. The feasibility of control will depend not only on scientific considerations but also on the financial and public health resources of the community, or country.

The Roll Back Malaria (RBM) Partnership was launched in 1998, and is a joint initiative of WHO, UNDP, UNICEF and the World Bank, initially concentrated on Africa. Since then the partnership has expanded to incorporate non-government agencies and academic institutions, and its implementation has been extended to other parts of the world such as Asia and Latin America. It advocates an integrated approach to malaria control, involving impregnated mosquito nets, indoor house-spraying, community-level drug treatment and, if available, vaccines. The goal is a 50% reduction in malaria mortality by the year 2010, and a 75% reduction by 2015.

Further reading

Asidi, A. N., N'Guessan, R., Koffi, A. A. *et al.* (2005) Experimental hut evaluation of bednets treated with an organophosphate (chlorpyrifos-methyl) or a pyrethroid (lambdacyalothrin) alone and in combination against insecticide-resistant *Anopheles gambiae. Malaria Journal* (online), **4**, 25.

Beier, J. C. (1998) Malaria parasite development in mosquitoes. *Annual Review of Entomology*, **43**, 519–43.

Casman, E. A. and Dowlatabadi, H. (eds.) (2002) *The Contextual Determinants of Malaria*. Washington, DC: RFF Press.

Chavasse, D., Reed, C. and Attawell, K. (eds.) (1999) *Insecticide Treated Net Projects: a Handbook for Managers*. London and Liverpool: Malaria Consortium.

Collins, F. H. and Paskewitz, S. M. (1995) Malaria: current and future prospects for control. *Annual Review of Entomology*, **40**, 195–219.

Coluzzi, M. and Bradley, D. (1999) The malaria challenge after one hundred years of malariology. *Parassitologia*, **41**, (1–3), 1–528.

Curtis, C. F., Maxwell, C. A., Magesa, S. M., Rwegoshora, T. and Wilkes, T. J. (2006) Insecticide-treated bed-nets for malaria mosquito control. *Journal of the American Mosquito Control Association*, **22**, 501–6.

Curtis, C. F. and Townson, H. (1998) Malaria: existing methods of vector control and molecular entomology. *British Medical Bulletin*, **54**, 311–25.

Dean, M. (2001) *Lymphatic Filariasis: the Quest to Eliminate a 4000-Year-Old Disease*. Hollis, NH: Hollis Publishing Company.

Dobson, M., Malowany, M. and Stapleton, D. (eds.) (2000) Dealing with malaria in the last sixty years. Proceedings of a Rockefeller Foundation Conference held in New York, May 11–14, 1998. *Parassitologia*, **42** (1–2), 1–182.

Gilles, H. M. and Warrell, D. A. (eds.) (2002) *Essential Malariology*, 4th edn. London: Edward Arnold.

Harbach, R. E. (2004) The classification of the genus *Anopheles* (Diptera: Culicidae): a working hypothesis of phylogenic relationships. *Bulletin of Entomological Research*, **94**, 537–53.

Hill, J., Lines, J. and Rowland, M. (2006). Insecticide-treated nets. In *Control of Human Parasitic Diseases*, ed. D. H. Molyneux. *Advances in Parasitology*, **61**, London: Academic Press, pp. 77–128.

Lengeler, C. (2004) Insecticide-treated bednets and curtains for preventing malaria. *Cochrane Database of Systematic Reviews* (2004) **2**, CD000363.

Lindblade, K. A., Gimnig, J. E., Kamau, L. *et al.* (2006) Impact of sustained use of insecticide-treated bednets on malaria vector species distribution and culicine mosquitoes. *Journal of Medical Entomology*, **43**, 428–32.

Litsios, S. (1996) *The Tomorrow of Malaria*. Wellington, New Zealand: Pacific Press.

Nájera, J. A. (2001) Malaria control: achievements, problems and strategies. *Parassitologia*, **43** (1–2), 1–98.

Nájera, J. A., Kouznetsov, R. L. and Delacollette, C. (1998) *Malaria Epidemics: Detection and Control, Forecasting and Prevention*. WHO/MAL/98.1084. Geneva: World Health Organization.

Roll Back Malaria Partnership (2005) *Scaling-up Insecticide-Treated Netting Programmes in Africa: a Strategic Framework for Coordinated National Action*. Revision 12, 2nd edn. Geneva: WHO.

Sasa, M. (1976) *Human Filariasis: a Global Survey of Epidemiology and Control.* Tokyo: University of Tokyo Press.

Sharma, S. K., Upadhyay, A. K., Haque, M. A. *et al.* (2006) Wash resistant and bioefficacy of Olyset nets: a long-lasting insecticide-treated mosquito net against malaria vectors and nontarget household pests. *Journal of Medical Entomology*, **43**, 884–8.

Spielman, A. (2006) Ethical dilemmas in malaria control. *Journal of Vector Control*, **31**, 1–8.

Wernsdorfer, W. H. and McGregor, I. (eds.) (1988) *Malaria: Principles and Practice of Malariology.* Volumes 1 and 2. Edinburgh: Churchill Livingstone.

World Health Organization (1989) *Geographical Distribution of Arthropod-Borne Diseases and their Principal Vectors.* WHO/VBC/89.967. Geneva: WHO Vector Biology and Control Division.

World Health Organization (1992) Lymphatic filariasis: the disease and its control. *World Health Organization Technical Report Series*, **821**, 1–71.

World Health Organization (1992) *Entomological Field Techniques for Malaria Control. Part 1: Learner's Guide.* Geneva: WHO.

World Health Organization (1992) *Entomological Field Techniques for Malaria Control. Part II: Tutor's Guide.* Geneva: WHO.

World Health Organization (1993) *A Global Strategy for Malaria Control.* Geneva: WHO.

World Health Organization (1993) Implementation of the global malaria control strategy. *World Health Organization Technical Report Series*, **839**, 1–57.

World Health Organization (1994) *Lymphatic Filariasis Infection and Disease: Control Strategies.* Report of a consultative meeting at the Universitii Sains Malaysia Penang, Malaysia (August 1994). TDR/CTD/FIL/PENANG/94.1, 1–30.

World Health Organization (1995) Vector control for malaria and other mosquito-borne diseases. *World Health Organization Technical Report Series*, **857**, 1–91.

Zahar, A. R. (1984–96) A series of ten World Health Organization mimeographed documents entitled 'Vector bionomics in the epidemiology and control of malaria', covering Europe, Africa, the southern and eastern Mediterranean regions, south-western Arabia, Asia west of India, South-east Asia and the western Pacific region.

See also references at the ends of Chapters 1 and 3.

Learning Resources
Centre

3

Culicine mosquitoes (Culicinae)

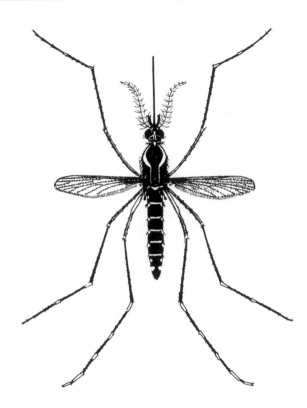

The subfamily Culicinae contains 38 genera, but as already mentioned (see Chapter 1) some taxonomists recognize many more genera, two of which are *Stegomyia* and *Ochlerotatus*. However, in this book all species attributed to these two genera are retained in the genus *Aedes*, with *Stegomyia* and *Ochlerotatus* recognized as subgenera.

The medically most important genera are *Culex*, *Aedes*, *Haemagogus*, *Sabethes* and *Mansonia*, while *Coquillettidia* and *Psorophora* are of lesser importance. Species of *Culex*, *Aedes* and *Coquillettidia* are found in both temperate and tropical regions, whereas *Psorophora* species occur only in North, Central and South America. *Haemagogus* and *Sabethes* mosquitoes are restricted to Central and South America. *Mansonia* is mainly tropical.

Certain *Aedes* mosquitoes are vectors of yellow fever in Africa, and *Aedes*, *Haemagogus* and *Sabethes* are yellow fever vectors in Central and South America. *Aedes* species are also vectors of the classical and haemorrhagic forms of dengue. All six genera of culicine mosquitoes mentioned here, as well as some others, can transmit a variety of other arboviruses. Some *Culex*, *Aedes* and *Mansonia* species are important vectors of filariasis (*Wuchereria bancrofti* or *Brugia malayi*). *Psorophora* species are mainly pest mosquitoes but a few transmit arboviruses, while the *Coquillettidia* species *Cq. crassipes* is one of the vectors of brugian filariasis.

Characters separating the subfamily Culicinae from the Anophelinae have been outlined in Chapter 1 and are summarized in Table 1.1.

It is not easy to give a reliable and non-technical guide to the identification of the most important culicine genera. Nevertheless, characters that will usually separate these genera are given below, together with notes on their biology.

3.1 *Culex* mosquitoes

3.1.1 Distribution
Culex mosquitoes are found more or less worldwide, but they are absent from the extreme northern parts of the temperate zones.

3.1.2 Eggs
Eggs are usually brown, long and cylindrical, laid upright on the water surface and placed together to form an ***egg raft*** which can comprise up to about 300 eggs (Fig. 1.15). No glue or cement-like substance binds the eggs to each other; adhesion is due to surface forces holding the eggs together. Eggs of a few other mosquitoes, including those of the genus *Coquillettidia*, also deposit their eggs in rafts.

3.1.3 Larvae
The larval siphon is often long and narrow (Fig. 3.1), but it may be short and fat. There is always ***more than one pair*** of ***subventral tufts*** of hairs on

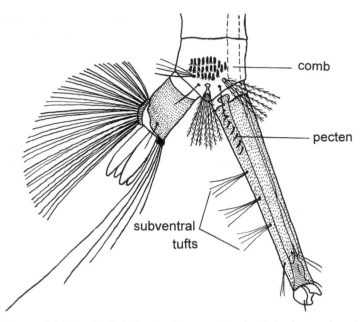

Figure 3.1 Terminal abdominal segments of a *Culex* larva, showing the long siphon with more than one (three shown here) subventral tufts of hair.

the siphon, none of which is near the base. These hair tufts may consist of very few short and simple hairs, which can be missed unless larvae are carefully examined under a microscope.

3.1.4   Adults
Frequently, but not always, the thorax, legs and wing veins of the adult are covered with ***dull-coloured***, often brown, scales (Plate 2). The abdomen is often covered with brown or blackish scales, but some whitish scales may occur on most segments. Adults are recognized more by their lack of ornamentation than by any striking diagnostic characters. The tip of the female abdomen is not pointed but blunt. Claws on all tarsi are simple and those on the hind tarsi are very small. Examination under a microscope shows that all tarsi have a pair of small fleshy ***pulvilli*** (Fig. 1.2).

3.1.5   Biology
Eggs are laid in a great variety of aquatic habitats. Most *Culex* species breed in ***ground collections*** of water such as pools, puddles, ditches, borrow pits and ricefields. Some lay their eggs in man-made container-habitats such as tin cans, water receptacles, bottles and water-storage tanks. Only a few species breed in tree-holes and even fewer in leaf axils. The medically most important species, *Culex quinquefasciatus*, which is a vector of bancroftian

filariasis, breeds in waters polluted with organic debris such as rotting vegetation, household refuse and excreta. Larvae of this species are commonly found in partially blocked drains and ditches, soakaway pits and septic tanks, and in village pots, especially abandoned ones in which water is polluted and unfit for drinking. It is a mosquito that is associated with urbanization, especially towns with poor and inadequate drainage and sanitation. Under these conditions its population increases rapidly. Adults mainly bite at night.

Culex tritaeniorhynchus is an important vector of Japanese encephalitis and breeds in ricefields and grassy pools. In southern Asia larvae are not uncommon in fish ponds which have had manure added to them.

Culex quinquefasciatus, and many other *Culex* species, bite humans and other hosts *at night*. Some species, such as *C. quinquefasciatus*, commonly rest indoors both before and after feeding, but they also shelter in outdoor resting places.

3.2   *Aedes* mosquitoes

3.2.1   Distribution
Worldwide, the range of *Aedes* mosquitoes extends well into northern and arctic areas, where they can be vicious biters and serious pests to people and livestock.

3.2.2   Eggs
Eggs are usually black, more or less ovoid in shape, and are always laid *singly* (Fig. 1.15). Careful examination shows that the eggshell has a distinctive mosaic pattern. Eggs are laid on damp substrates just beyond the water line, such as on damp mud and leaf litter of pools, on the damp walls of clay pots, rock-pools and tree-holes.

Eggs can withstand *desiccation*, the intensity and duration of which varies, but in many species they can remain dry but viable for many months. When flooded, some eggs may hatch within a few minutes, while others of the same batch may require longer immersion in water, so that hatching may be spread over several days or weeks. Even when eggs are soaked for long periods some may fail to hatch because they require several soakings followed by short periods of desiccation before hatching can be induced. Even if environmental conditions are favourable, eggs may be in a state of diapause and will not hatch until this resting period is terminated. Various stimuli, including reduction in the oxygen content of water, changes in daylength, and temperature, may be required to break diapause in *Aedes* eggs.

Many *Aedes* species breed in small *container-habitats* (tree-holes, plant axils, etc.) which are susceptible to drying out; thus the ability of eggs to withstand *desiccation* is clearly advantageous. Desiccation and the ability

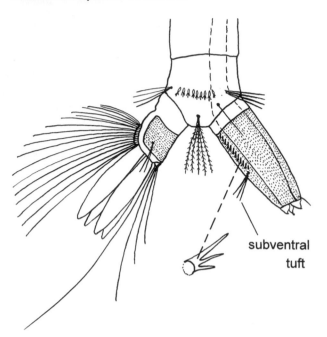

Figure 3.2 Terminal abdominal segments of an *Aedes* larva, showing the short siphon with a single subventral hair tuft.

of eggs to hatch in instalments can create problems with controlling the immature stages (see p. 77).

3.2.3   Larvae

Aedes species usually have a short barrel-shaped siphon, and there is only *one pair* of *subventral tufts* (Fig. 3.2) which never arises from less than one-quarter of the distance from the base of the siphon. There are at least three pairs of setae in the ventral brush, the antennae are not greatly flattened and there are no very large setae on the thorax. These characters should separate *Aedes* larvae from most of the culicine genera, but not unfortunately from larvae of South American *Haemagogus*. In Central and South America *Aedes* larvae can usually be distinguished from those of *Haemagogus* by possessing either larger or more strongly spiculate antennae and the comb not being on a sclerotized plate as in some *Haemagogus*.

3.2.4   Adults

Many, but not all, *Aedes* adults have conspicuous *patterns* on the thorax formed by black, white or silvery scales (Fig. 3.3, Plate 3); in some species yellow and/or brownish scales are present. The legs often have dark and white rings (Fig. 3.4b). *Aedes aegypti*, often called the yellow fever mosquito,

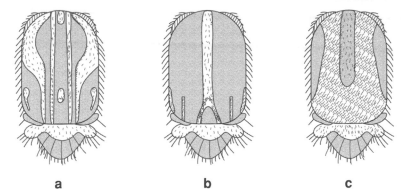

Figure 3.3 Dorsal surfaces of the thoraces of adult *Aedes* mosquitoes, showing examples of thoracic patterns of dark and pale scales: (a) *Aedes aegypti* with black scales and typical lyre-shaped silvery markings; (b) *Aedes albopictus* with prominent silvery central line; (c) *Aedes trivittatus* with dark brown and whitish, not silvery, scales.

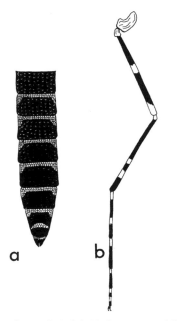

Figure 3.4 (a) Abdomen and (b) leg of an *Aedes* adult, showing typical arrangement of black and white scales.

is readily recognized by the lyre-shaped silver markings on the lateral edges of the scutum (Fig. 3.3a). Scales on the wing veins of *Aedes* mosquitoes are narrow and usually mainly black. In *Aedes* the abdomen is often covered with black and white scales forming distinctive patterns, and in the female the abdomen is pointed at its tip (Fig. 3.4a).

3.2.5   Biology

Although some *Aedes* species breed in marshes and ground pools, including snow-melt pools in arctic and subarctic areas, many, especially tropical species, are found in natural **container-habitats** such as tree-holes,
bamboo stumps, leaf axils and rock-pools, or in man-made ones such as
water-storage pots, tin cans and tyres. For example, *Ae. aegypti* breeds in
water-srorage pots or jars that are either inside or outside houses. Larvae
occur mainly in those having clean water intended for drinking. In some
areas *Ae. aegypti* also breeds in rock-pools and tree-holes. *Aedes africanus*,
an African species involved in the sylvatic transmission of yellow fever,
breeds mainly in tree-holes and bamboo stumps, whereas *Ae. bromeliae*,
another African yellow fever vector, breeds almost exclusively in leaf axils,
especially those of banana plants, pineapples and coco-yams (*Colocasia*).

Aedes albopictus, a vector of dengue in South-east Asia, breeds in both
natural and man-made container-habitats such as tree-holes, water-storage
pots and vehicle tyres. This species was introduced into the USA in 1985
as dry but viable eggs that had been oviposited in tyres in Asia and then
exported. It can also be introduced into countries by eggs in lucky bamboo (*Dracaena* species), in which there is an increasing trade. By 2006 *Ae.
albopictus* had spread to more than 29 states in the USA. It is now found in
many Latin American countries, in some sub-Saharan African countries, in
15 European countries, in Israel and in both Australia and New Zealand.
In summary it has been reported from more than 26 countries outside Asia.
However, *Ae. albopictus* has often not managed to establish itself in countries with more temperate climates or where efficient control has rapidly
eliminated invasions.

Larvae of *Ae. polynesiensis* occur in man-made and natural containers,
especially split coconut shells, whereas larvae of *Ae. pseudoscutellaris* are
found in tree-holes and bamboo stumps. Both species are important vectors
of diurnally subperiodic bancroftian filariasis. *Aedes togoi*, a minor vector of
nocturnally periodic bancroftian and brugian filariasis, breeds principally
in rock-pools containing fresh or brackish water.

The life cycle of *Aedes* mosquitoes from eggs to adults can be rapid, taking
as little as seven days, but more usually it takes 10–12 days; in temperate
species the life cycle may last several weeks to many months, and some
species overwinter as eggs or larvae.

Adults of most *Aedes* species bite mainly during the **day** or early evening.
Most biting occurs out of doors and adults usually rest out of doors before
and after feeding.

3.3   *Haemagogus* mosquitoes

3.3.1   Distribution

Haemagogus mosquitoes are found only in Central and South America.

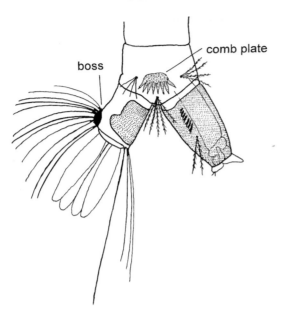

Figure 3.5 Terminal abdominal segments of a *Haemagogus* larva, showing a ventral brush arising from a dark sclerotized boss, and comb scales arranged on a small 'plate'.

3.3.2 Eggs

Eggs are usually black and ovoid and laid singly in tree-holes and other natural container-habitats, and occasionally in man-made ones. There is no simple method of distinguishing eggs of *Haemagogus* from those of *Aedes* or *Psorophora* mosquitoes.

3.3.3 Larvae

Larvae have only one pair of subventral tufts arising, as in *Aedes*, not less than a quarter of the distance from the base of the siphon. They resemble *Aedes* larvae but can usually be distinguished by the antennae being short and either lacking spicules or with just a very few, and by a ventral brush arising from a sclerotized **boss** (Fig. 3.5). In some species the comb teeth are arranged at the edge of a sclerotized **plate**; in *Aedes* this plate is absent.

3.3.4 Adults

Adults are very **colourful** and can easily be recognized by the presence of broad, flat and bright metallic blue, red, green or golden scales, covering the dorsal part of the thorax. Like *Sabethes* mosquitoes (Fig. 3.7a) they have exceptionally large **antepronotal thoracic lobes** behind the head. *Haemagogus* adults are rather similar to *Sabethes*, and it may be difficult for the novice to separate these two genera. However, no *Haemagogus* mosquito has paddles on the legs, which is a conspicuous feature of many, but not all, *Sabethes* species (Fig. 3.7c).

3.3.5   Biology

Eggs can withstand *desiccation*. Larvae occur mostly in tree-holes and bamboo stumps, but also in rock-pools, split coconut shells and sometimes in assorted domestic containers. They are basically *forest mosquitoes*. Adults bite during the day but mostly in the tree-tops, where they feed on monkeys. Under certain environmental conditions, however, such as experienced at edges of forests during tree-felling operations or during the dry season, they may descend to the forest floor to bite humans and other hosts. Species such as *Haemagogus spegazzinii*, *Hg. leucocelaenus* and *Hg. janthinomys* are all involved in yellow fever transmission in forested areas.

3.4   *Sabethes* mosquitoes

3.4.1   Distribution

Sabethes mosquitoes are found only in Central and South America.

3.4.2   Eggs

Little is known about the eggs of *Sabethes* species, but they are laid singly, have no prominent surface features such as bosses or sculpturing and are incapable of withstanding desiccation. Eggs of *Sabethes chloropterus*, a species sometimes involved in the sylvatic cycle of yellow fever, are rhomboid in shape and can thus be readily identified from most other culicine eggs (Fig. 3.7b).

3.4.3   Larvae

The larval siphon is relatively slender and moderately long (Fig. 3.6) and has many *single* hairs placed ventrally, laterally or dorsally. *Sabethes* larvae can usually be distinguished from other mosquito larvae by having only *one pair* of setae in the ventral brush, the comb teeth arranged in a single row, or at most with three or four detached teeth, and by the absence of a pecten.

3.4.4   Adults

The dorsal surface of the thorax is covered with appressed *iridescent* blue, green and red scales (Plate 4). The *antepronotal lobes*, like those in *Haemagogus*, are very large (Fig. 3.7a). Adults of many species have one or more pairs of tarsi with conspicuous *paddles* composed of narrow scales (Fig. 3.7c). Their presence immediately distinguishes *Sabethes* from all other mosquitoes. Species which lack these paddles resemble those of *Haemagogus*, and a specialist is required to distinguish them.

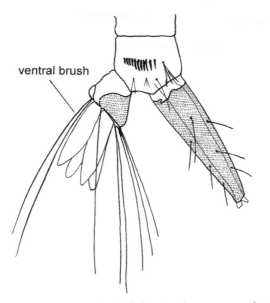

ventral brush

Figure 3.6 Terminal abdominal segments of a *Sabethes* larva, showing a single pair of hairs in the ventral brush, numerous single hairs on the siphon, and absence of a pecten on the siphon.

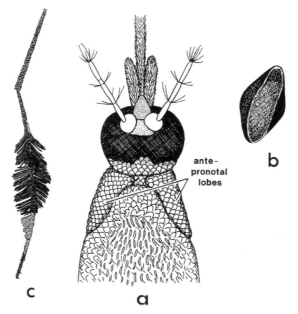

ante–
pronotal
lobes

b

c

a

Figure 3.7 *Sabethes* mosquitoes: (a) head and thorax, showing antepronotal lobes forming a 'collar' behind the head; (b) an egg; (c) hind-leg, showing long narrow scales forming a 'paddle'.

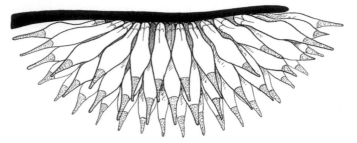

Figure 3.8 *Mansonia* eggs glued to the underside of floating aquatic vegetation.

3.4.5   Biology
Larvae occur in tree-holes and bamboo stumps; a few species are found in leaf axils of bromeliads and other plants. They are *forest mosquitoes*. Adults bite during the day and mainly in the tree canopy, but like *Haemagogus* adults may descend to ground level at certain times to bite humans and other hosts. *Sabethes chloropterus* has been incriminated as a sylvan vector of yellow fever.

3.5   *Mansonia* mosquitoes

3.5.1   Distribution
Mansonia is principally a genus of wet tropical areas, but a very few species occur in temperate regions.

3.5.2   Eggs
Eggs are dark brown or blackish and cylindrical, but have a tube-like extension apically which is usually darker than the rest of the egg. Eggs are laid in sticky compact *masses*, often arranged as a rosette, which are glued to the undersurface of floating vegetation (Figs. 1.15, 3.8).

3.5.3   Larvae
Mansonia larvae are easily recognized because they have *specialized siphons* adapted for piercing aquatic plants (Fig. 3.9b,c) to obtain air. The siphon tends to be conical with the apical part darker and heavily sclerotized, and it has teeth and curved hairs which enable a larva to attach to plants and insert its siphon. The pupal respiratory trumpets are also inserted into plants for respiration (Fig. 3.9a).

3.5.4   Adults
Typically adults have the legs (Fig. 3.10c), palps, wings and body covered with a *mixture* of dark (usually brown) and pale (usually white or creamy) scales, giving the mosquito a rather dusty appearance. The scattering of

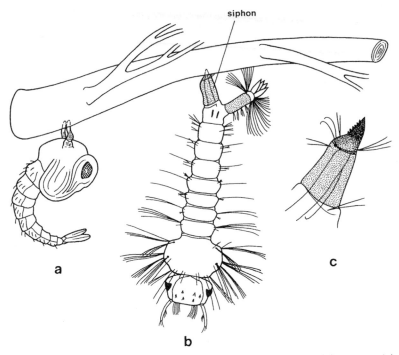

Figure 3.9 Immature stages of *Mansonia* mosquitoes: (a) pupa with respiratory trumpets inserted into an aquatic plant; (b) larva with siphon inserted into an aquatic plant for respiration; (c) larval siphon, showing serrated structures used to pierce aquatic plants.

dark and pale scales on the wing veins gives the wings the appearance of having been sprinkled with salt and pepper (Fig. 3.10a), and provides a useful character for identification. Closer examination shows that the scales on the wings are very broad and often asymmetrical and almost *heart-shaped* (Fig. 3.10b). In other mosquitoes these scales are longer and narrower.

3.5.5 Biology

Eggs are glued to the undersurface of plants and hatch within a few days; they are unable to withstand desiccation. All larval habitats have aquatic vegetation, either rooted (such as grasses, rushes and reeds) or floating (such as *Pistia stratiotes*, *Salvinia* or *Eichhornia* species). Larvae consequently occur in permanent collections of waters, such as swamps, marshes, ponds, borrow pits, grassy ditches, irrigation canals and even in the middle of rivers if they have floating plants.

Larvae and pupae only detach themselves from plants and rise to the water surface if they are disturbed. Because they are more or less permanently attached to plants the immature stages are frequently missed in

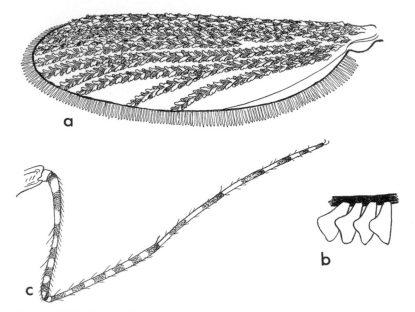

Figure 3.10 *Mansonia* mosquitoes: (a) wing, showing speckled distribution of dark and pale scales; (b) a few scales on a wing vein, showing their broad, almost heart shape; (c) leg, showing distribution of dark and pale scales giving a banded pattern.

larval surveys. It is therefore not easy to identify breeding places unless special collecting procedures are undertaken, such as removing plants and examining them for attached *Mansonia* larvae and pupae. Because it is difficult to get insecticides to the larvae, which may be some distance below the water surface, it is often difficult to control *Mansonia* mosquitoes with conventional insecticidal applications. However, herbicides can be used to kill the plants (see p. 79).

Adults usually bite at night, but a few species are day-biters. After feeding, most *Mansonia* rest out of doors, but a few species rest indoors. The main medical importance of *Mansonia* mosquitoes is as vectors of filariasis, such as the nocturnally periodic and nocturnally subperiodic forms of *Brugia malayi* in Asia. In Africa filariasis (*W. bancrofti*) is not transmitted by *Mansonia* species, although they can be vectors of a few, but not very important, arboviruses.

3.6   *Coquillettidia* mosquitoes

The genus *Coquillettidia* is of lesser medical importance and therefore described only briefly. It is mainly tropical but a few species occur in temperate regions. *Coquillettidia* is related to *Mansonia* and has sometimes been treated as a subgenus of *Mansonia*, but species of *Coquillettidia* differ in several respects. For example, as in *Culex* species, eggs are formed into *egg rafts* that float on the water, but they are narrower and longer than *Culex* rafts. Larvae have rather conical siphons, which like those of *Mansonia* are inserted into plants for respiratory purposes, but the larval antennae are much longer than those in *Mansonia*. Adults have narrow scales on the wings, not broad or heart-shaped ones as found in *Mansonia*, and several *Coquillettidia* species are a bright *yellow*. *Coquillettidia crassipes* can be a vector of nocturnally subperiodic *B. malayi* in South-east Asia.

3.7 *Psorophora* mosquitoes

Psorophora mosquitoes are of lesser medical importance and so are described briefly. They are found from Canada to South America. They are similar in many respects to *Aedes* species: for example their eggs look like *Aedes* eggs and like them they can withstand *desiccation*. Adults of some pest species are large mosquitoes. A specialist is required to distinguish the larvae from those of *Aedes* species. Breeding places are mainly flooded pastures and sometimes ricefields; larvae of several species are predators. Although they can be vectors of a few arboviruses, such as Venezuelan equine encephalitis, their main importance is as vicious biters.

3.8 Medical importance

3.8.1 Biting nuisance

A considerable amount of money is spent on mosquito control, not because they are vectors of disease but because they are troublesome biters. For example, some of the best-organized mosquito control operations are in North America, where more money may be spent on mosquito control than in most tropical countries, where mosquitoes are important vectors of disease. In northern temperate and subarctic areas of America, Europe and Asia much greater numbers of mosquitoes can be encountered biting people than in tropical countries. Although they may not be transmitting diseases to humans in these areas they can, nevertheless, make life outdoors intolerable.

3.8.2 Arboviruses

Numerous arboviruses are transmitted by culicine mosquitoes, including important ones such as those causing yellow fever and dengue. *Aedes* and *Culex* species are the most important vectors of arboviruses.

The intrinsic incubation period of a virus in humans is usually just a few days, often 3–4 days, after which virus appears in the peripheral blood and the host is viraemic. *Viraemia* lasts only a few days, typically 3 days, after which the virus disappears from the peripheral blood. A vector must bite a viraemic host if it is to become infected and transmit an infection.

A relatively high titre of arbovirus is usually required before a virus can pass across the gut-wall of the mosquito into the haemolymph. From the haemolymph the virus invades many tissues and organs, including the *salivary glands*, where virus multiplication occurs. This is the *extrinsic incubation period* of development and can take 5–30 days, depending on temperature, the type of virus and the mosquito species. In most mosquito-borne viral infections the extrinsic cycle is typically 8–15 days.

3.8.3   Yellow fever (*Flavivirus*)

The arbovirus causing yellow fever occurs in Africa and tropical areas of the Americas. It does not occur in Asia or elsewhere, although mosquitoes capable of transmitting the disease occur in many countries. Yellow fever is a *zoonosis*, being essentially an infection of forest monkeys which under certain conditions can be transmitted to humans.

Africa

In Africa the yellow fever virus occurs in certain cercopithecid monkeys (e.g *Colobus* and *Cercopithecus* species) inhabiting the forests. Other primates also act as reservoir hosts, and in East Africa the lesser bushbaby (*Galago senegalensis*) is an important host. The virus is transmitted amongst these primates by tree-hole-breeding mosquitoes, mainly *Aedes africanus*. This forest-dwelling mosquito bites mainly in the forest canopy soon after sunset – just in the right place and at the right time to bite monkeys going to sleep in the tree-tops. This *sylvatic*, savannah, forest or monkey cycle, as it is sometimes called, maintains a virus reservoir in the monkey population (Fig. 3.11). In Africa, monkeys are little affected by yellow fever, dying only occasionally; but in East Africa infected bushbabies die. Some of the monkeys involved in the forest cycle descend from the trees to steal bananas from farms at the edge of the forest. In this habitat the monkeys get bitten by different mosquitoes. For example, by *Aedes bromeliae* (formerly called *Ae. simpsoni*), a species that breeds in leaf axils of bananas, plantains, coco-yams (*Colocasia* species) and pineapples, and in West Africa also by other species such as tree-hole-breeding *Aedes furcifer*, *Ae. taylori* and *Ae. luteocephalus*. These species bite during the day at the edges of forests, and if the monkeys have viraemia, that is yellow fever virus circulating in their peripheral blood, the mosquitoes become infected. If these mosquitoes live sufficiently long they can transmit yellow fever to other monkeys, or more importantly to people. This transmission cycle, occurring in clearings at the edge of forests involving both monkeys and humans, is sometimes referred to as the *rural cycle* (Fig. 3.11). When people return to their villages, or travel to towns, they get bitten by different mosquitoes, including *Ae. aegypti*, a domestic species breeding mainly in man-made containers such as water-storage pots, abandoned tin cans and vehicle tyres. If people have viraemia then *Ae. aegypti* becomes infected and yellow fever is transmitted among the human population by this species. This is the *urban cycle* of yellow fever transmission (Fig. 3.11).

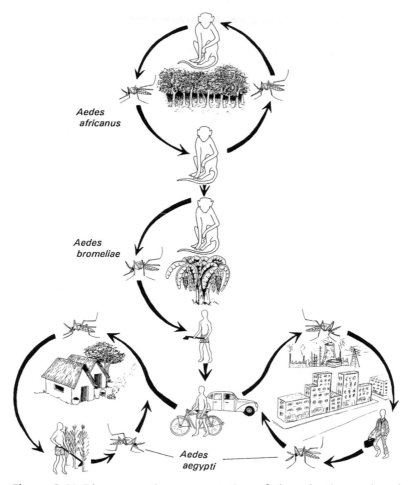

Figure 3.11 Diagrammatic representation of the sylvatic, rural and urban cycles of yellow fever transmission in Africa (only major vectors included).

The epidemiology of yellow fever is complicated and variable. In some areas, for example, yellow fever may be circulating among the monkey population yet rarely gets transmitted to humans because local vector mosquitoes are predominantly zoophagic.

Yellow fever virus may be *transovarially* transmitted in *Aedes* species. That is eggs and subsequent generations arising from infected adults are born already infected with virus. (Transovarial transmission is more usually associated with tick-borne diseases than with mosquitoes.) The virus can also be passed from congenitally infected males to females during mating, which in effect is venereal transmission. How important these routes of infection are remains unclear, but such mechanisms, especially transovarial transmission, would provide a means of virus survival during long dry seasons.

Americas

In Central and South America the yellow fever cycle, although similar to that in Africa, differs in certain aspects (Fig. 3.12). As in Africa it is an infection of forest monkeys, mainly cebid ones (e.g. howler, *Alouatta* species, and spider monkeys, *Ateles* species), and is transmitted among them by forest-dwelling mosquitoes. In the sylvatic or jungle cycle the vectors are *Haemagogus* species including *Hg. spegazzinii*, *Hg. janthinomys* and *Hg. leucocelaenus* and to a lesser extent *Sabethes chloropterus*, and sometimes *Aedes* species. These are all arboreal mosquitoes which bite in the forest canopy and breed in tree-holes or bamboo. New World monkeys are more susceptible to yellow fever than African monkeys and frequently become sick and die. When people enter the jungle to cut down trees for timber, mosquitoes, which normally bite monkeys at canopy heights, may descend and bite them; if these mosquitoes are infected people will develop yellow fever. The disease is then spread from person to person in villages and towns, as in Africa, by *Ae. aegypti* (Fig. 3.12) and this constitutes the **urban cycle**, but this has not been documented in South America since 1942.

Transovarial transmission has been reported in *Haemagogus* species.

3.8.4   Dengue (*Flavivirus*)

A number of different viruses (dengue types 1, 2, 3 and 4) are responsible for dengue. Dengue is widely distributed in the tropics, occurring throughout most of South-east Asia, the Pacific, the Indian subcontinent, sub-Saharan Africa, Central and South America and the Caribbean. Dengue haemorrhagic fever causes infant mortality and has appeared in many parts of South-east Asia and India. In 1981 haemorrhagic dengue and dengue type 4 were first noticed in Cuba. Since then in the Americas dengue type 4 has been reported from at least 26 countries and haemorrhagic dengue from 18 countries. Dengue is now the most important viral disease transmitted by mosquitoes, having been recorded from more than 100 countries, and dengue has increased exponentially in the last few decades. Both dengue and haemorrhagic dengue are transmitted by *Ae. aegypti*, and in South-east Asia to a lesser extent also by *Ae. albopictus*; both species breed in natural and man-made container-habitats such as water-storage pots and tyres. Mosquitoes of the *Ae. scutellaris* group, which also breed in natural and man-made containers, may also be vectors in the Pacific islands. Although transmission of dengue virus amongst monkeys in forests of Malaysia, Vietnam and West Africa has been reported, there is little evidence that enzootic strains are involved in epidemics. Transmission is among the human population, and humans are the reservoir hosts.

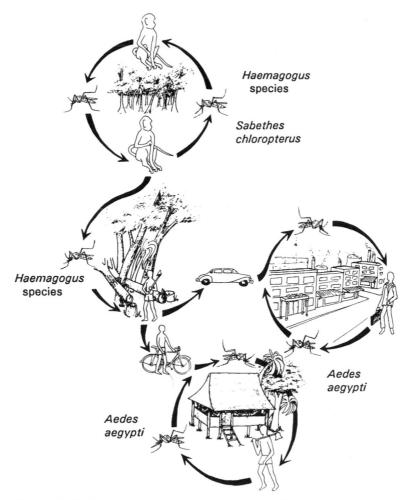

Figure 3.12 Diagrammatic representation of the jungle, rural and urban cycles of yellow fever transmission in Central and South America.

3.8.5 West Nile virus (WNV) (*Flavivirus*)

West Nile virus is a member of the Japanese encephalitis group. It originated from Uganda and has now spread to the Middle East, many African countries and 24 European countries. In 1999 the virus was recorded for the first time from the Americas, in New York State. How the virus entered North America is unresolved, but by 2006 it had been recorded in 50 US states and six Canadian provinces, as well as in Mexico.

The virus is principally an infection of birds. Crows (*Corvus* species) are the commonest birds in the USA to be found dead and infected with WNV, but many other bird species are infected and laboratory experiments

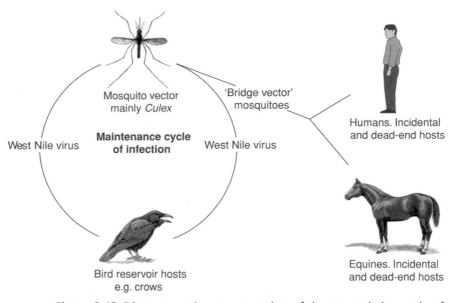

Mosquito vector mainly *Culex*

'Bridge vector' mosquitoes

West Nile virus

Maintenance cycle of infection

West Nile virus

Humans. Incidental and dead-end hosts

Bird reservoir hosts e.g. crows

Equines. Incidental and dead-end hosts

Figure 3.13 Diagrammatic representation of the transmission cycle of West Nile virus.

show some to be potentially more efficient hosts. Virus has been isolated from more than 67 mosquito species, although *Culex* mosquitoes, such as the *Cx. pipiens* complex, *Cx. modestus* and *Cx. univittatus* are among the most important vectors. (It has been shown that there can be co-feeding transmission: when an uninfected mosquito is feeding on a host very near an infected mosquito the virus from the infected mosquito passes to the uninfected mosquito, thus making it a potential vector.) Occasionally a mosquito species that feeds on both birds and mammals, a so-called bridge vector, transfers the infection to humans, horses and other mammals. Mammals are incidental hosts and are termed **dead-end hosts** because mosquitoes feeding on infected mammals cannot pick up sufficient virus to infect further mammals when they bite them (Fig. 3.13).

3.8.6   Japanese encephalitis (JE) (*Flavivirus*)

This virus is found throughout most of Asia from the Indian subcontinent to Japan and much of South-east Asia. The basic transmission cycle involves mosquitoes biting water birds, mainly herons, egrets and ibises, which are the principal reservoir hosts. Some infected mosquitoes bite pigs, which develop high viraemia and are termed **amplifying hosts**. If infected mosquitoes, having bitten birds or pigs, then bite humans they transmit the virus to them. Humans, however, are dead-end hosts and consequently there is no human-to-human transmission (Fig. 3.14). Transmission to birds, humans and pigs is mainly by *Culex tritaeniorhynchus*, *Cx. vishnui* and

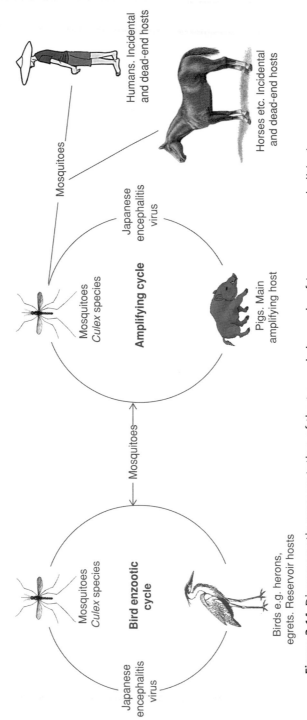

Figure 3.14 Diagrammatic representation of the transmission cycle of Japanese encephalitis virus.

Cx. pseudovishnui, all of which are ricefield-breeding mosquitoes. *Culex gelidus*, breeding in streams and ricefields, is another vector and probably maintains the virus in pig-to-pig transmissions. In southern India *Mansonia indiana* is a secondary vector.

3.8.7   Other arboviruses

There are many other mosquito-borne arboviruses infecting humans in various parts of the world, for example Ross River virus (*Alphavirus*: Australasia), Sindbis (*Alphavirus*: Africa, Asia, Australia, Europe), Murray Valley encephalitis (*Alphavirus*: Australia, Papua New Guinea) and Rift Valley fever (*Bunyavirus*: Africa, Saudi Arabia, Yemen). These and the more important arboviral diseases very briefly discussed below are **zoonoses**, as of course is yellow fever.

Chikungunya (CHIK) (*Alphavirus*)

Found in sub-Saharan Africa, the Indian subcontinent and throughout much of South-east Asia. In 2006 large outbreaks occurred in several islands in the Indian Ocean such as Réunion, Mayotte, Mauritius, Madagascar and the Seychelles, and as a result many infected tourists returned home to Europe and North America. In sub-Saharan Africa there are about five principal vectors, all being *Aedes* species such as *Ae. africanus*; however, *Ae. aegypti* is not usually an important vector, whereas in Asia it is the main vector. Transmission in the Indian Ocean islands is by *Ae. aegypti* and *Ae. albopictus*. In Africa the virus is possibly maintained in non-human primates, but in Asia it is maintained in a human-to-human cycle. In 2007 transmission of CHIK virus was reported for the first time in Europe, in Italy.

St Louis encephalitis (SLE) (*Flavivirus*)

In southern Canada and widely distributed in the USA, extending into Central and South America. In eastern USA it is mainly an urban disease spread by *Cx. quinquefasciatus* or *Cx. pipiens*. **Amplifying hosts** are chickens and peridomestic wild birds. In western USA SLE occurs mainly in rural areas with the ricefield-breeding *Cx. tarsalis* as a vector.

Eastern equine encephalitis (EEE) (*Alphavirus*)

In Canada, but mainly in eastern USA, and extending down to South America. It is probably the most severe encephalitis virus in humans and horses. EEE is principally an infection of birds, and in North America is spread among them mainly by *Culiseta melanura* (the genus *Culiseta* is not discussed in this book). It is transmitted to people and horses by various species of *Aedes*, *Culex* and *Coquillettidia*.

Western equine encephalitis (WEE) (*Alphavirus*)

In western USA and extending into South America. This is basically an arboviral infection of birds, which are **amplifying** and **maintenance** hosts.

Transmission is by *Cx. tarsalis*, a ricefield-breeding mosquito, as well as by other *Culex* and *Aedes* species. *Aedes* mosquitoes also transmit the virus to mammals, including humans and horses.

Venezuelan equine encephalitis (VEE) (*Alphavirus*)

In southern USA through Central America to northern parts of South America. The infection is often fatal in horses but usually very mild in humans. Rodents are believed to be important *amplifying hosts* involved in enzootic transmission cycles, and birds and bats may be involved in dispersing the virus. Vectors are mainly *Culex* species. Epizootic/epidemic transmission involves equines, which are the principal amplifying hosts. *Aedes taeniorhynchus*, *Culex taeniopus* and other *Culex* species in the subgenus *Melaniconion*, as well as *Psorophora confinnis*, are important vectors.

Viraemic titres produced in humans by JE, EEE, SLE, VEE and sometimes also by WEE are so low that the infection cannot be transmitted by mosquitoes from humans to humans or from humans to other susceptible hosts. Humans are thus *dead-end hosts* for these viruses, as are horses for the encephalitis viruses infecting them.

3.8.8　Filariasis

The development in mosquitoes of filarial worms causing lymphatic filariasis is briefly described in Chapter 2 in connection with the role of anopheline vectors. Both bancroftian and brugian filariasis occur in two distinct forms. The *nocturnally periodic* form, in which the microfilariae are in the peripheral blood only at night, is transmitted by night-biting mosquitoes such as *Anopheles*, *Mansonia* and *Culex quinquefasciatus*. During the day the microfilariae are in the blood vessels supplying the lungs, and are not available to be taken up by mosquitoes. In *subperiodic* forms of *Wuchereria bancrofti* and *Brugia malayi* the microfilariae exhibit a reduced periodicity and are present in the peripheral blood during the day as well as at night, but there nevertheless remains a degree of periodicity. For example, subperiodic *W. bancrofti*, such as found in Polynesia, has a small peak in microfilarial density during the daytime and can therefore be called *diurnally subperiodic*, whereas subperiodic *B. malayi* in West Malaysia, Sumatra, Sabah, Thailand, etc. has a slight peak of microfilariae at night, and so can be called *nocturnally subperiodic*.

The list of vectors of lymphatic filariasis is complex, with some 40 *Anopheles* species and 40 culicine species belonging to four genera being incriminated as vectors of bancroftian and brugian filariasis. Only some of the more important species can be given in the following account and in Table 3.1. For a more detailed account of the vectors see the table in Zagaria and Savioli (2002).

Table 3.1. *Summary of principal mosquito vectors of filariasis*

Species and forms of filariasis	Geographical distribution	Vectors	Zoonotic reservoir hosts
Wuchereria bancrofti			
Nocturnally periodic	Throughout tropics (except Polynesia)	*Anopheles, Culex quinquefasciatus*	None
	Papua New Guinea	*Anopheles, Culex annulirostris Mansonia uniformis*	
	Philippines	*Aedes poicilius, Anopheles*	
Diurnally subperiodic	Polynesia in general	*Aedes polynesiensis*	None
	Fiji	*Aedes pseudoscutellaris*	
	New Caledonia	*Aedes vigilax*	
Nocturnally subperiodic	Thailand	*Aedes niveus*	None
Brugia malayi			
Nocturnally periodic (principally open swamps)	South Asia, from India eastwards	*Anopheles, Mansonia annulata, Ma. annulifera, Ma. uniformis*	Not important, possibly some exist
	China	*Aedes togoi*	
Nocturnally subperiodic (mainly in swampy forests)	Malaysia, Indonesia, Thailand, Philippines	*Mansonia bonneae, Ma. dives, Coquillettidia crassipes*	Monkeys, especially leaf-monkeys (*Presbytis* species), wild and domestic cats, pangolins
Brugia timori			
Nocturnally periodic	Alor, Timor, Flores, other Indonesian islands	*Anopheles barbirostris*	None

Bancroftian filariasis

Wuchereria bancrofti occurs throughout much of the tropics and subtropics including South America, sub-Saharan Africa, Asia and the South Pacific. It is the most widely distributed filarial infection of humans. Bancroftian filariasis is mainly an urban disease. There are no animal reservoir hosts and the parasites develop only in mosquitoes and humans.

The ***nocturnally periodic*** form is transmitted in Asia, South America and East Africa by *Culex quinquefasciatus*. This mosquito is widespread in the tropics and breeds mainly in waters polluted with human or animal faeces or rotting vegetation and other filth; larvae are found in septic tanks, cesspits, pit latrines, drains and ditches, and in water-storage jars if they contain organically polluted water. This mosquito has increased in numbers in many towns due to increasing urbanization and the proliferation of insanitary collections of water. Adults bite at night, and after feeding they often rest in houses. Although *Cx. quinquefasciatus* is an efficient vector in much of Africa, it is a poor vector in West Africa, where most transmission of bancroftian filariasis is by *Anopheles gambiae* and *An. funestus* and the disease is principally rural. *Anopheles* species are also vectors in parts of Asia, while in Papua New Guinea *Anopheles* are the main vectors although *Mansonia uniformis* and *Culex annulirostris* are also important in rural areas.

In the Philippines *Aedes poicilius* is the most important vector. Adults bite in the early part of the night, mainly indoors but also sometimes outdoors. After feeding, adults rest outdoors. Larvae occur in leaf axils of banana, plantain and coco-yam (*Colocasia*) plants.

The ***diurnally subperiodic*** form occurs in the Polynesian region, from where the nocturnally periodic form is absent. The most important vector is *Aedes polynesiensis*, a day-biting mosquito which feeds mostly outdoors but may enter houses to feed; adults rest almost exclusively out of doors. Larvae occur in natural containers such as split coconut shells, leaf bracts and crab-holes, and also in man-made containers such as discarded tins, pots, vehicle tyres and canoes. *Aedes pseudoscutellaris* is another outdoor day-biting mosquito that is a vector of diurnally subperiodic *W. bancrofti*, especially in Fiji. It breeds mainly in tree-holes and bamboo stumps but larvae are also found in crab-holes. In New Caledonia *Ae. polynesiensis* is absent and the most important vector is *Ae. vigilax*, adults of which feed outdoors mainly during the day. Larvae are found in brackish or fresh water in rock-pools and ground pools.

The ***nocturnally subperiodic*** form is found in Thailand and is transmitted by the *Aedes niveus* group of mosquitoes. Adults bite and rest outdoors; larvae are found mainly in bamboo.

It should be noted that although several *Aedes* mosquitoes are vectors of filariasis, especially the bancroftian form, *Ae. aegypti* is not a vector of lymphatic filariasis. Infection rates of mosquitoes with infective larvae of *W. bancrofti* range usually from about 0.1% to 5%, depending greatly on vector species and local conditions. But in Singapore infection rates of 20% have been recorded, and in East Africa 30%.

Brugian filariasis

The **nocturnally periodic** form is principally a rural disease. It occurs throughout most of Asia, from southern India to Malaysia, Vietnam, Cambodia, Thailand and Indonesia. It is transmitted by night-biting mosquitoes, such as *Anopheles* species (Chapter 2) and *Mansonia* species such as *Ma. uniformis*, and in parts of India also by *Ma. annulifera*. *Mansonia* species breed in more or less permanent waters such as swamps and ponds having floating or rooted aquatic vegetation; in Kerala, India, *Ma. annulifera* also breeds in coconut soakage pits. Adults bite mainly outdoors and rest out of doors after feeding, but they may bite and rest indoors in some areas. There are no known important animal reservoirs of the nocturnally periodic form.

The **nocturnally subperiodic** form of *B. malayi* occurs in west Malaysia, Indonesia, Thailand and the Philippines, and is transmitted by *Mansonia* mosquitoes such as *Ma. dives, Ma. bonneae* and *Ma. annulifera*, and in Thailand and the Philippines also by *Coquillettidia crassipes*. Larvae of all these species occur in habitats with much vegetation, such as swampy forests. Adults bite mainly at night although *Ma. dives* and *Ma. bonneae* may also bite during the day. This subperiodic form of *B. malayi* is a **zoonosis** and essentially a parasite of swamp monkeys, especially the so-called leaf monkeys (*Presbytis* species). Humans become infected when they live at the edges of these areas. Other reservoir hosts include *Macaca* monkeys, domestic and wild cats, such as civets, and pangolins (i.e. scaly anteaters, genus *Manis*).

Infection rates of mosquitoes with infective larvae of *B. malayi* range from about 0.1% to 2–3%, which is slightly lower than for *W. bancrofti*, but infection rates vary according to mosquitoes and local conditions.

As mentioned in Chapter 2, the discovery of filarial worms in mosquitoes does not necessarily imply they are vectors of either *W. bancrofti* or *B. malayi*, because mosquitoes are also vectors of several other filarial parasites of animals.

3.9   Control

Repellents, mosquito nets and mosquito screening of houses and other personal protection measures (discussed in Chapters 1 and 2) can give some relief from culicine mosquitoes. It is, however, more difficult to obtain protection from culicines than from anophelines because many of them bite outdoors during the daytime. Spraying the interior surfaces of houses with residual insecticides, as practised for *Anopheles* control, is not usually effective for culicines, because most species do not rest in houses. Control is aimed mainly at the larvae, although aerial ultra-low-volume (ULV) applications are sometimes used to kill adult culicine mosquitoes.

3.9.1 *Aedes* and *Psorophora*

Control of mosquitoes such as *Ae. aegypti*, *Ae. polynesiensis* and *Ae. albopictus*, species which breed mainly in man-made containers in both rural and urban areas, is often aimed at reducing the numbers of larval habitats, that is control by **source reduction**. Thus, people are encouraged not to store water in pots or allow water to accumulate in discarded tin cans, bottles, vehicle tyres, etc. However, persuading people to cooperate in reducing peridomestic breeding of mosquitoes is often difficult unless local legislation is strictly enforced. A reliable piped water supply to houses can do much to reduce *Ae. aegypti* breeding.

When source reduction is not feasible insecticides can be used. Although insecticidal spraying can kill *Aedes* and *Psorophora* larvae it usually has no effect on their dry, but viable, eggs. These have been deposited at the edges of larval habitats and will hatch when the water level rises and floods them. *Aedes* and *Psorophora* mosquitoes breeding in pools, ponds and marshy areas can be controlled by ground-based or aerial applications of granular organophosphate insecticides, or insect growth regulators. Applications can be made either before or after habitats have become flooded, that is **pre-** or **post-flood treatments**. Insecticidal granules landing on dry or muddy grounds remain more or less inactive until the habitats become flooded. When this occurs previously dry eggs hatch, but at the same time flooding causes the release of insecticide from the granules and this kills the newly hatched larvae. This technique helps overcome problems of controlling *Aedes* and *Psorophora* mosquitoes, whose eggs may hatch in instalments over extended periods after flooding.

Ground-based or aerial ULV application of malathion, fenitrothion, pirimiphos-methyl or pyrethroids, mainly to kill adult mosquitoes, is often the most appropriate control strategy in epidemic situations.

Insecticides used to kill mosquito larvae in water intended for drinking must have extremely low mammalian toxicity; they should also impart no taste to the water. The insecticide usually recommended is temephos, in the form of 1% in briquettes or sand granules that will give a concentration of 1 mg active ingredient per litre of water. However, some communities refuse to have their potable water dosed with any insecticide (see page 28).

Yellow fever and dengue

Africa has over 90% of the world's yellow fever cases. The best defence against yellow fever is to use the attenuated 17D vaccine. Vector control, through sustained reduction in mosquito breeding, nevertheless still has a role in reducing the risks of yellow fever outbreaks; it also remains the main approach for dengue control, because there are presently no vaccines approved for widespread use. Epidemics of dengue, and sometimes also yellow fever, can be curtailed by killing the adult vectors by aerial ULV insecticidal spraying, where the main objective is to kill infected adult mosquitoes as rapidly as possible.

3.9.2   *Culex*

Culex quinquefasciatus, an important filariasis vector, is best controlled by improving sanitation and installing modern sewage systems, but often this is not feasible and insecticidal measures have to be employed. In most areas *Cx. quinquefasciatus* is resistant to a wide range of insecticides and this limits the choices of chemicals that can be used. Larval habitats should be sprayed every 7–10 days, and usually relatively large dosage rates are needed because most insecticides are less effective in the presence of organic pollution, which is characteristic of *Cx. quinquefasciatus* breeding places. Chlorpyrifos is one of the more effective insecticides in polluted waters. Insect growth regulators such as methoprene, diflubenzuron and pyriproxyfen have also been used against *Cx. quinquefasciatus*.

Tipping non-toxic expanded *polystyrene beads* (2–3 mm) into pit latrines and cesspits to completely cover the water surface with a 2–3 cm layer suffocates larvae and pupae as well as preventing female *Cx. quinquefasciatus* from ovipositing. A single application can persist for several years and give excellent control. This is a control method that is readily accepted by most communities.

Insecticidal house-spraying, as practised against malaria vectors, and insecticide-impregnated bed-nets can be effective against *Cx. quinquefasciatus* and other *Culex* species if they are both endophilic and night-biters.

In North America ULV spraying has frequently been used against vectors of the encephalitis viruses.

Lymphatic filariasis

In 1998 the Global Programme to Eliminate Lymphatic Filariasis (GPELF) was initiated, with the aim of achieving this goal by 2020.

Control strategies involve mass distribution of microfilarial drugs administered annually, using albendazole and ivermectin in sub-Saharan Africa, and elsewhere albendazole plus diethylcarbamazine, and concurrently reducing vector populations. Application of suitable larvicides, including the microbial insecticides *Bacillus thuringiensis* var. *israelensis* and *B. sphaericus*, use of expanded polystyrene beads in pit latrines and septic tanks, use of insecticide-impregnated bed-nets, and possibly residual house-spraying, can all reduce densities of *Cx. quinquefasciatus*, a very important filariasis vector. In contrast it is very difficult to reduce populations of *Mansonia* mosquitoes, important filariasis vectors in Asia. This is because adults rarely rest or bite indoors and the larval breeding sites are often large and relatively inaccessible.

In 2006–7 Egypt and China reported the interruption of transmission, with no new filarial disease cases occurring.

3.9.3 *Mansonia*

Mansonia mosquitoes are usually controlled either by removing aquatic weeds upon which the larvae and pupae depend for their oxygen requirements, or by using herbicides to kill the weeds. Removing weeds, however, may result in ecological changes that allow aquatic habitats to become colonized by mosquito species that were previously excluded by the dense covering of weeds.

Insecticidal granules or pellets are more suitable than liquid formulations for killing *Mansonia* larvae because they penetrate vegetation, sink to the bottom of larval habitats and release their insecticidal contents through the water. However, species such as *Ma. dives* and *Ma. bonneae*, which are important vectors of brugian filariasis, breed in extensive swampy inaccessible forests where control is impractical.

Further reading

Curtis, C. F., Malecela-Lazaro, M., Reuben, R. and Maxwell, C. A. (2002) Use of floating layers of polystyrene beads to control populations of the filaria vector *Culex quinquefasciatus. Annals of Tropical Medicine and Parasitology*, **96** (suppl. 2), 97–104.

Barrett, A. D. T. and Higgs, S. (2007) Yellow fever: a disease that has yet to be conquered. *Annual Review of Entomology*, **52**, 209–29.

Diallo, M., Ba, Y., Sall, A. A. *et al.* (2003) Amplification of the sylvatic cycle of dengue virus type 2, Senegal, 1999–2000: entomologic findings and entomologic considerations. *Emerging Infectious Diseases*, **9**, 362–7.

Gratz, N. (2006) *Vector- and Rodent-borne Diseases in Europe and North America*. Cambridge: Cambridge University Press.

Gratz, N. and Knudsen, A. B. (1996) *The Rise and Spread of Dengue, Dengue Haemorrhagic Fever and its Vectors: a Historical Review (up to 1995)*. CTD/FIL(DEN) 96.7. Geneva: World Health Organization.

Gubler, D. J. and Kuno, G. (eds.) (1997) *Dengue and Dengue Haemorrhagic Fever*. Wallingford: CAB International.

Halstead, S. B. and Gomez-Dantes, H. (eds.) (1992) *Dengue: a World-Wide Problem, a Common Strategy*. Proceedings of the international conference on dengue and *Aedes aegypti* community-based control. Mexico: Ministry of Health, Rockefeller Foundation.

Komar, N., Langevin, S., Hinten, S. *et al.* (2003). Experimental infection of North American birds with the New York 1999 strain of West Nile virus. *Emerging Infectious Diseases*, **9**, 11–22.

Monath, T. P. (ed.) (1988) *Arboviruses: Epidemiology and Ecology. Volume 1: General Principles. Volume 2: African Horse Sickness to Dengue. Volume 3: Eastern Equine Encephalomyelitis to O'nyong nyong. Volume 4: Oropouche Fever to Venezuelan Equine Encephalomyelitis. Volume 5: Vesicular Stomatitis to Yellow Fever*. Boca Raton, FL: CRC Press.

Monath, T. P. (2001) Yellow fever. In *The Encyclopedia of Arthropod-transmitted Infections of Man and Domesticated Animals*, ed. M. W. Service. Wallingford: CABI, pp. 571–7.

Muller, R. (2002) *Worms and Human Diseases*, 2nd edn. Wallingford: CABI.

Mutebi, J. P. and Barrett, A. D. T. (2002) The epidemiology of yellow fever in Africa. *Microbes and Infections*, **4**, 1459–68.

Ottesen, E. A. (2003) Lymphatic filariasis: treatment, control and elimination. *Advances in Parasitology*, **61**, 1–47.

Reeves, W. C. (ed.) (1990) *Epidemiology and Control of Mosquito-Borne Arboviruses in California, 1943–1987*. Sacramento, CA: California Mosquito and Vector Control Association.

Tomori, O. (1999) Impact of yellow fever on the developing world. *Advances in Virus Research*, **53**, 5–34.

Weaver, S. C., Ferro, C. Barrera, R., Boshell, J. and Navarro, J.-C. (2004) Venezuelan equine encephalitis. *Annual Review of Entomology*, **49**, 141–74.

White, G. B. and Nathan, M. B. (eds.) (2002) The elimination of lymphatic filariasis: public-health challenges and the role of vector control. *Annals of Tropical Medicine and Parasitology*, **96** (suppl. 2), 1–164.

World Health Organization (1997) *Dengue Haemorrhagic Fever: Diagnosis, Treatment, Prevention and Control*, 2nd edn. Geneva: WHO.

World Health Organization (2005) Global programme to eliminate lymphatic filariasis. *Weekly Epidemiological Record*, **80**, 202–12.

Zagaria, N. and Savioli, L. (2002) Elimination of lymphatic filariasis: a public-health challenge. *Annals of Tropical Medicine and Parasitology*, **96** (suppl. 2), 3–13.

See also references at the ends of Chapters 1 and 2.

4

Black-flies (Simuliidae)

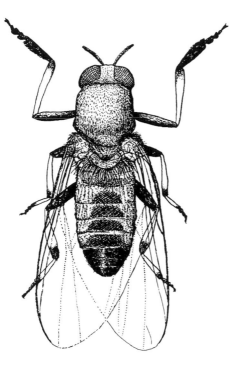

Black-flies belong to the family Simuliidae and have a worldwide distribution. There are approaching 2000 species in about 26 genera. However, only four genera – *Simulium, Prosimulium, Austrosimulium* and *Cnephia* – contain species that bite people.

Medically, *Simulium* is by far the most important genus as it contains many vectors. In Africa, species in the *S. damnosum* complex and the *S. neavei* group, and in Central and South America, species in the *S. ochraceum*, *S. metallicum* and *S. exiguum* complexes, transmit the parasitic nematode *Onchocerca volvulus*, which causes human onchocerciasis (river blindness). In Brazil, *S. amazonicum* transmits *Mansonella ozzardi*, a filarial parasite that is usually regarded as non-pathogenic.

4.1   External morphology

The Simuliidae are commonly known as black-flies, but in some areas, especially Australia, they may be called sand-flies. As explained in Chapter 5, this latter terminology is confusing and best avoided because biting flies in the family Ceratopogonidae are also sometimes called sand-flies, while flies in the subfamily Phlebotominae are regarded as the true sand-flies.

Adult black-flies are quite small, about 1.5–4 mm long, relatively stout-bodied and, when viewed from the side, have a rather *humped* thorax. As their vernacular name indicates they are usually *black* in colour (Plate 5) but some species have contrasting patterns of white, silvery or yellowish hairs on their bodies and legs, while others may be predominantly orange or bright yellow.

Black-flies have a pair of large compound eyes, which in females are separated on the top of the head (a condition known as *dichoptic*) whereas in males the two eyes touch each other and occupy most of the head (a condition known as *holoptic*). In males, but not females, the lenses of the eyes are larger on the upper half than on the lower half (Fig. 4.1b). The antennae are short, stout, cylindrical and distinctly segmented (usually 11 segments) and lack long hairs. The mouthparts are short and relatively inconspicuous, but the five-segmented maxillary palps hang downwards and are easily seen. *Only females* bite. The mouthparts, being short and broad, do not penetrate deeply into the host's tissues. Teeth on the labrum stretch the skin, while the rasp-like action of the maxillae and mandibles cuts through the skin and ruptures its fine blood capillaries. The small pool of blood produced is then sucked up by the fly. This method of feeding is ideally suited for picking up the microfilariae of *Onchocerca volvulus*, which occur in human skin, not blood. Morphologically the mouthparts are similar to those of the biting midges (Ceratopogonidae, Chapter 6).

The thorax is covered dorsally with very fine and appressed hairs, which can be black, white, silvery, yellow or orange and may be arranged in various patterns. The relatively short legs are also covered with very fine

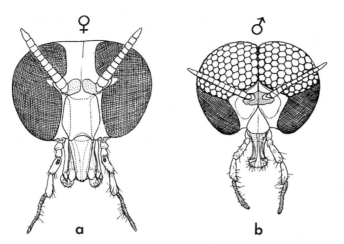

Figure 4.1 Front view of adult *Simulium* heads; (a) female with dichoptic eyes; (b) male with holoptic eyes.

and closely appressed hairs and may be unicolorous or have contrasting pale and dark bands. Each tarsus has a pair of claws, which are untoothed (i.e. simple) in mammal-feeders.

Wings are characteristically short and broad and *lack* both scales and prominent hairs. Only the veins near the anterior margin are well developed; the rest of the wing is membranous and has indistinct venation (Fig. 4.2, Plate 5). The wings are colourless or almost so, and when at rest are closed over the body like the blades of a closed pair of scissors.

The abdomen is short and squat, and covered with inconspicuous closely appressed fine hairs. In neither sex are the genitalia very conspicuous. Black-flies are easily *sexed* by looking to see whether their eyes are dichoptic (females) or holoptic (males).

4.2   Life cycle

Eggs are about 0.1–0.4 mm long, brown or black, and are more or less triangular in shape but have rounded corners and smooth unsculptured shells (Fig. 4.3a); they are covered with a sticky substance. Eggs are always laid in *flowing water*, but the type of breeding place differs greatly according to species. Habitats can vary from small trickles of water, slow-flowing streams, lake outlets and water flowing from dams to fast-flowing rivers and rapids. Some species prefer lowland streams and rivers whereas others are found in mountain rivers. In species such as *S. ochraceum*, a Central American vector of onchocerciasis, eggs are scattered over the surface of flowing water while females are in flight. In most species, however, ovipositing females alight on partially immersed objects such as rocks, stones and vegetation to lay some 150–800 eggs in sticky masses or strings.

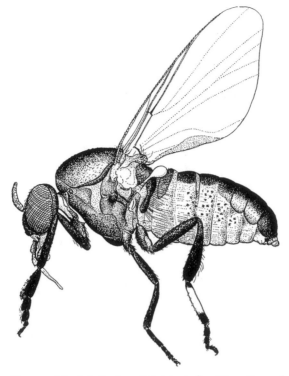

Figure 4.2 Adult simuliid black-fly (*Simulium damnosum*) in lateral view. (Courtesy of R. W. Crosskey and the Natural History Museum, London.)

Females may crawl underneath the water and become completely submerged during oviposition. There are sometimes a few favoured oviposition sites in a stream or river, resulting in thousands of eggs from many females being found together. *Simulium damnosum*, for example, frequently has such communal oviposition sites. Eggs are unable to survive desiccation.

Eggs of *S. damnosum* hatch within about 1–2 days, but in many other tropical species the egg stage lasts 2–4 days. Eggs of species inhabiting temperate or cold northern areas may not hatch for many weeks, and some species pass the winter as diapausing eggs.

There are six to eleven (usually seven) larval instars and the mature larva, depending on the species, is about 4–12 mm long. It is easily distinguished from all other aquatic larvae (Fig. 4.3b). The head is black, or almost so, and has a prominent pair of *feeding brushes* (cephalic fans), while the weakly segmented, cylindrical body is usually greyish, but may be darker or sometimes even greenish. The body is slightly swollen beyond

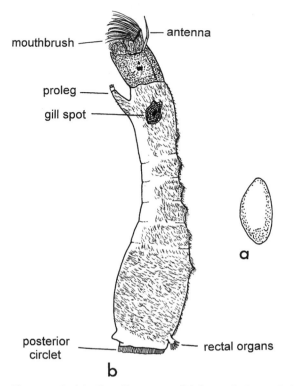

mouthbrush

antenna

proleg

gill spot

a

posterior circlet

rectal organs

b

Figure 4.3 (a) *Simulium* egg; (b) lateral view of the last larval instar showing principal diagnostic simuliid characters. The species figured is *Simulium damnosum*, a species which has the body covered with minute dark setae and has small dorsal tubercles, i.e. small humps.

the head and in most, but not all, species distinctly swollen towards its end. Ventrally, just below the head, is a small thoracic ***proleg*** which is armed with a small circlet of hooklets. The rectum has finger-like rectal organs which on larval preservation may be extruded and visible as a dorsal protuberance towards the end of the abdomen.

Larvae do not swim but remain sedentary for long periods on submerged vegetation, rocks, stones and other debris. Attachment is achieved by the ***posterior hook-circlet*** (anal sucker of many previous authors) tightly gripping a small silken pad which has been produced by the larva's very large salivary glands and which is firmly glued to the substrate. Larvae can nevertheless move about. This is achieved by alternately attaching themselves to the substrate by the proleg and the posterior hook-circlet, which results in their moving in a looping manner. When larvae are disturbed they can deposit sticky saliva on a submerged object, release their hold and be swept downstream for some distance at the end of a silken thread. They can then either swallow the thread of saliva and regain their original position, or re-attach themselves at a site further downstream.

Larvae normally orientate themselves to lie *parallel* to the flow of water with their heads downstream. They are mainly filter-feeders, ingesting, with the aid of large feeding brushes, suspended particles of food. However, a few species have predaceous larvae and others are occasionally cannibalistic. Depending on species and temperature larval development may be as short as 6–12 days, but in some species it may be extended to several months, and in other species larvae overwinter.

Mature larvae can be recognized by a blackish mark termed a *gill spot* (respiratory organ of the future pupa) on each side of the thorax (Fig. 4.3b). These larvae spin, with the silk produced by their salivary glands, a protective slipper-shaped brownish *cocoon*. This is firmly attached to submerged vegetation, rocks or other objects, and its shape and structure vary greatly according to species (Fig. 4.4). After weaving the cocoon the enclosed larva pupates. The pupa has a pair of thin-walled *respiratory gills*, which are usually prominent and may be filamentous or broad. Their length, shape and number of filaments or branches provide useful taxonomic characters for species identification. These gills, and the anterior part of the pupa, often project from the entrance of the cocoon (Fig. 4.4). In both tropical and non-tropical countries the pupal period lasts only 2–6 days and is unusual in not appearing to be dependent on temperature. On emergence adults either rise rapidly to the water surface in a protective bubble of gas, which prevents them from being wetted, or they escape by crawling up partially submerged objects such as vegetation or rocks. A characteristic of many species is the almost simultaneous mass emergence of thousands of adults. On reaching the water surface the adults immediately take flight.

The empty pupal cases, with gill filaments still attached, remain enclosed in their cocoons after the adults have emerged and provide a means of identifying simuliid species that have successfully emerged.

A few African and Asian black-fly species have a very unusual aquatic association. For example, in East Africa larvae and pupae of *S. neavei* do not occur on submerged rocks or vegetation but on other aquatic arthropods, such as on the immature stages (nymphs) of mayflies (Ephemeroptera), and various crustaceans including freshwater crabs. Such an association is termed a *phoretic* relationship. As eggs are not found on these animals they are probably laid on submerged stones or vegetation.

Nuclei of the larval salivary gland cells have large *polytene chromosomes* which have banding patterns that are used to identify otherwise morphologically identical species within a species complex. For example, chromosomal studies have shown that there are 55 cytoforms in the *S. damnosum* complex, many of which are known to be distinct species and have scientific names.

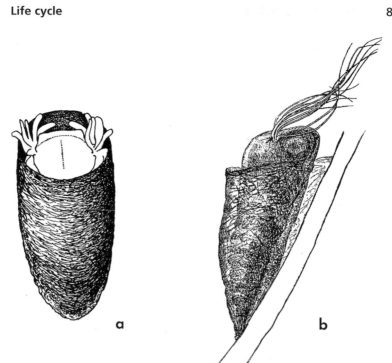

Figure 4.4 *Simulium* pupae in their cocoons: (a) dorsal view of a species (*S. damnosum*) with broad and short respiratory filaments (courtesy of R. W. Crosskey and the Natural History Museum, London); (b) lateral view of a species with long and thin respiratory filaments.

4.2.1   Adult behaviour

Both male and female black-flies feed on plant juices and naturally occurring sugary substances, but *only females* take blood-meals. Biting occurs out of doors during the day. Many species, including *S. damnosum* in Africa, have bimodal biting patterns, with peak biting in the early morning and again in the afternoon or early evening. However, in some species, such as *S. ochraceum* in Guatemala, biting continues more or less throughout the day. Many species seem particularly active on cloudy, overcast days and in thundery weather. Black-flies may exhibit marked preferences for feeding on different parts of the body; for example, *S. damnosum* feeds mainly on the legs whereas *S. ochraceum* prefers to bite the head and torso. When feeding on animals, adults crawl down the fur of mammals, or feathers of birds, to bite the host's skin; they may also enter the ears to feed.

Many species of black-fly feed almost exclusively on birds (*ornithophagic*), others on non-human mammalian hosts (*zoophagic*), while several species also bite people (*anthropophagic*). Some of these, however, prefer various large animals such as donkeys or cattle and bite humans only as a second choice, whereas others find humans almost equally attractive hosts; no species feeds exclusively on people. After feeding, blood-engorged females shelter in vegetation, on trees and in other natural outdoor resting places until the blood-meal is completely digested. In the tropics this takes 2–3 days, while in non-tropical areas it may take 3–8 days or longer, the speed of digestion depending mainly on temperature. A few species can lay eggs without a blood-meal (i.e. they are *autogenous*). Relatively little is known about black-fly longevity, but it seems that adults of most species live for 3–4 weeks.

Females of some species may fly considerable *distances* (15–30 km) from their emergence sites to obtain blood-meals. They may also be dispersed long distances by winds. For example, it is not exceptional for adults of *S. damnosum* to be found biting 60–100 km from their breeding places, and in West Africa there is evidence that prevailing winds can carry adults up to 400–600 km. These long distances can hinder control programmes, because areas freed from black-flies can be reinvaded from distant breeding places. In Central and South America black-flies generally disperse only about 2–15 km.

In temperate and northern areas of the Palaearctic and Nearctic regions biting nuisance from simuliids is seasonal. This is because adults die in the autumn and new generations do not appear until the following spring or early summer. Although in many tropical areas there is continuous breeding throughout the year, there may nevertheless be dramatic increases in population size during the rainy season.

4.3   Medical importance

4.3.1   Annoyance
In both tropical and non-tropical regions black-flies can cause a very serious biting problem. Although the severity of the reaction to bites differs in different individuals, localized swelling and inflammation frequently occurs, accompanied by intense irritation lasting for several days, or even weeks. Repeated biting by black-flies such as *S. erythrocephalum* in central Europe, *S. posticatum* in England, *S. venustum* and *S. vittatum* in North America can cause headaches, fevers, swollen lymph glands and aching joints. In some areas of North America outdoor activities are almost impossible at certain times of the year due to the intolerable numbers of biting simuliids. The classical example of the nuisance caused by black-flies was the seasonal exodus during the eighteenth century of people from the Danube valley area in central Europe, largely to save their livestock from attack by enormous numbers of *S. colombaschense*. However, it is as disease vectors that black-flies are most important medically.

4.3.2   Onchocerciasis
Onchocerciasis is a non-fatal disease, often called river blindness, that is caused by the filarial parasite *Onchocerca volvulus*. There are no animal

reservoir hosts, so the disease is not a zoonosis. An estimated 17.6 million people are infected, of whom about 770 000 are blind or partially blind in 37 endemic countries. Onchocerciasis occurs throughout West Africa, Central Africa and much of East Africa from Ethiopia to Tanzania, with isolated pockets of infection in Malawi, Sudan and southern Yemen, possibly extending into Saudi Arabia. About 99% of all cases occur in 28 countries in Africa. In the Americas onchocerciasis is localized in areas of southern Mexico, Guatemala, Brazil, Venezuela, Ecuador and Colombia.

Black-flies are the only vectors of human onchocerciasis. Their habit during feeding of tearing and rasping the skin to rupture blood capillaries makes them particularly suited to the ingestion of the **skin-borne microfilariae** of *O. volvulus*. Most microfilariae ingested during feeding are destroyed or excreted, but some penetrate the stomach wall and migrate to the thoracic muscles. Here they develop into sausage-shaped stages (called L_1 larvae), then moult to L_2 stage larvae, a few of which moult again to become elongate and thinner L_3 worms which pass through the head and down the proboscis. These **infective third-stage** worms (about 660 μm in length) leave the proboscis and penetrate the host's skin when black-flies feed. The interval between the ingestion of microfilariae to the time infective larvae (L_3) are in the proboscis is about 6–12 days, depending on temperature.

African vectors of onchocerciasis

Chromosomal studies show that the *S. damnosum* complex is composed of about 55 cytoforms, many of which merit species status. The *damnosum* complex is widespread in tropical Africa and some of the species are the most important vectors of onchocerciasis.

Adults of the *damnosum* complex are mainly black. They can be recognized by their broad and flattened front tarsi with a conspicuous dorsal crest of fine hairs, and by a very broad white area on the first segment of the hind tarsus (basitarsus) (Fig. 4.2). Larvae are more or less covered with fine setae and there are usually prominent dorsal abdominal tubercles (Fig. 4.3b). Branches of the pupal respiratory gills are thick and finger-like and are located within the neck of the cocoon (Fig. 4.4a). The immature stages are found in the rapids of small or very large rivers in both savannah and forested areas of Africa. Adults frequently disperse hundreds of kilometres from their breeding sites. The vectorial status of species within the *damnosum* complex varies, the most important vectors of onchocerciasis being *S. damnosum* s. str., *S. sirbanum*, *S. sanctipauli* and *S. leonense*.

Other, but much less important, vectors include *S. neavei*, which is responsible for transmission in the Democratic Republic of Congo (formerly Zaire) and Uganda. *Simulium neavei* is a **phoretic** species: its larvae and pupae are found attached to other freshwater fauna, in this instance crabs of the genus *Potamonautes* and mayfly (Ephemeroptera) nymphs, both of which occur in small, rocky, rather turbid streams and rivers. *Simulium* *neavei* was eradicated from Kenya in the 1950s by bush clearing and insecticidal dosing of streams.

American vectors of onchocerciasis

Simulium ochraceum is widely distributed in Central America and northern parts of South America and is the principal onchocerciasis vector in southern Mexico and Guatemala. Adults are very small and are easily recognized by their dark brown legs, bright orange scutum (dorsal surface of the thorax) and yellow basal part of the abdomen, which contrasts with the black apical part. Females oviposit while in flight, dropping their eggs onto floating vegetation. Larval habitats consist of trickles of flowing water and very small streams, often concealed by bushes, vegetation and fallen leaves. Adults do not appear to disperse far. The main biting season is unusual in being in the drier months of the year.

Simulium metallicum occurs in Mexico through Central America to northern areas of South America. In northern Venezuela it appears to be the most important vector whereas in other areas such as in Mexico and Guatemala it is considered a minor vector. It is a black species and has a broad white area on the first segment of the hind tarsus. Larvae occur in small or large streams and rivers. Adults fly further from their breeding sites than do those of *S. ochraceum*.

Simulium exiguum is the only known vector in Colombia, and a primary vector in Ecuador. Above 150 m in the Brazilian Amazonas the main vector is *S. guianense*; below this height the vector is *S. oyapockense*.

4.3.3 *Mansonella ozzardi*

Mansonella ozzardi is a filarial parasite of humans that is usually regarded as non-pathogenic, although it has been reported as causing morbidity in Colombia and Brazil. It is transmitted in the Caribbean islands, Trinidad, Surinam and also Argentina by *Culicoides* species, mainly *C. furens* and *C. phlebotomus* (Chapter 6), but in northwestern Argentina, Brazil, Colombia, Guyana, Venezuela and southern Panama *S. amazonicum* is the main vector.

4.4   Control

Some protection can be gained by using repellents such as DEET, or by wearing pyrethroid-impregnated or sprayed clothing.

However, the only practical control method is to apply insecticides to larval breeding places. These need be applied to only a few selected sites on watercourses for some 15–30 minutes, because as the insecticide is carried downstream it kills simuliid larvae over long stretches of water. Flow rates of the water and its depth are used to calculate the quantity of insecticide to be released. In the past dosing rivers with DDT has given good control of ‚S. *damnosum* in Africa, but because of its accumulation in food chains DDT is no longer used. Nowadays temephos, phoxin, permethrin and etofenprox or *Bacillus thuringiensis* var. *israelensis* (*Bti*) are used. Treatment has to be regularly repeated, sometimes at 1–2 week intervals, throughout the year to prevent recolonization. In Guatemala insecticide-treated briquettes are used to control *S. ochraceum*, which breeds in small rivers.

In many areas ground application of larvicides is difficult, either because of the enormous size of the rivers requiring treatment or because breeding occurs in a large network of inaccessible small streams and watercourses. Under these conditions aerial applications from small aircraft or helicopters may be appropriate

4.4.1 Onchocerciasis control programme (OCP)

Because of the severity of river blindness in the Volta River Basin area of West Africa and its devastating effect on rural life, the world's most ambitious and largest vector control programme, the Onchocerciasis Control Programme (OCP), was initiated in 1974 by the World Health Organization. By 1986 there were 11 participating countries, namely Benin, Burkina Faso, Côte d'Ivoire, Ghana, Mali, Niger and Togo all receiving vector control, and Bissau, Guinea Bissau, Senegal and Sierra Leone without vector control but receiving community treatment with ivermectin. In the countries having vector control some 50 000 km of rivers over an area of 1.3 million km^2 that were breeding the *S. damnosum* complex were dosed weekly with temephos, which was dropped from helicopters or small aircraft. Because of the appearance of temephos resistance in 1980 in some populations and species of the *S. damnosum* complex, some rivers were treated with other insecticides or with *Bti*. To hinder the spread of further resistance a rotation of different insecticides was initiated in 1982. Larviciding continued in the countries added in 1986 (Guinea, Guinea Bissau, Senegal, Sierra Leone) until the year 2002.

Since 1988 the OCP has been undertaking large-scale distribution of the microfilaricidal drug *ivermectin* (Mectizan), which is given orally once or twice a year. Results from the OCP have been most spectacular, and transmission of river blindness has ceased over most of the OCP area.

In 1995 the African Programme for Onchocerciasis Control (APOC) was created to cover populations at risk in 19 countries outside the OCP. The objective was to establish, within 12 years, a sustainable community-based ivermectin treatment regimen, backed up with focal larviciding in some areas. Because ivermectin does not kill adult worms, control needs to continue for about 20–25 years to allow time for the reservoir of infection (adult onchocercal worms) in the human population to die out. There will be a phasing-out period of the programme (2008–10) and a final evaluation in 2010.

An estimated 95% of the population in Central and South America at risk of onchocerciasis live in Mexico, Guatemala and Venezuela, and in

1992 the Onchocerciasis Elimination Programme for the Americas (OEPA) was launched, based mainly on the distribution of ivermectin.

Further reading

Adler, P. H., Currie, D. C. and Wood, D. M. (2004) *The Black Flies (Simuliidae) of North America*. Ithaca, NY, and London: Comstock.

Anon. (1998) Mectizan and onchocerciasis: a decade of accomplishment and prospects for the future. The evolution of a drug into a development concept. *Annals of Tropical Medicine and Parasitology*, **92** (suppl. 1), 1–179.

Boatin, B. A. and Richards, F. O. (2006) Control of onchocerciasis. *Advances in Parasitology*, **61**, 349–54.

Crosskey, R. W. (1990) *The Natural History of Blackflies*. Chichester: Wiley.

Davies, J. B. (1994) Sixty years of onchocerciasis vector control: a chronological summary with comments on eradication, reinvasion, and insecticide resistance. *Annual Review of Entomology*, **39**, 23–45.

De Villiers, P. C. (1987) *Simulium* dermatitis in man: clinical and biological features in South Africa. *South African Medical Journal*, **71**, 523–5.

Hougard, J.-M., Yaméogo, L., Sékétéli, A., Boatin, B. and Dadzie, K. Y. (1997) Twenty-two years of black-fly control in the onchocerciasis control programme in West Africa. *Parasitology Today*, **13**, 425–8.

Molyneux, D. H. (2005) Onchocerciasis control and elimination: coming of age in resource-constrained health systems. *Trends in Parasitology*, **21**, 525–9.

Raybould, J. N. and White, G. B. (1979) The distribution, bionomics and control of onchocerciasis vectors (Diptera: Simuliidae) in eastern Africa and the Yemen. *Tropenmedizin und Parasitologie*, **30**, 505–47.

Service, M. W. (1977) Methods for sampling adult Simuliidae, with special reference to the *Simulium damnosum* complex. *Tropical Pest Bulletin*, **5**, 1–48.

Thylefors, B. and Allman, M. (2006) Towards the elimination of onchocerciasis. *Annals of Tropical Medicine and Parasitology*, **100**, 733–46.

World Health Organization (1997) *Twenty Years of Onchocerciasis Control: Review of the Work of the Onchocerciasis Control Programme in West Africa from 1974 to 1994*. Geneva: WHO. (In English or French.)

World Health Organization (2002) *Success in Africa: the Onchocerciasis Control Programme in West Africa, 1974–2002*. Geneva: WHO.

World Health Organization (2004) Onchocerciasis (river blindness): report from the thirteenth InterAmerican Conference on Onchocerciasis, Cartagena de Indias, Colombia. *Weekly Epidemiological Record*, **79**, 310–12.

5

Phlebotomine sand-flies
(Phlebotominae)

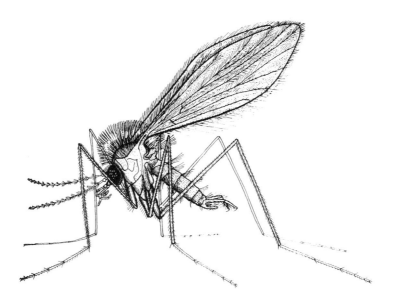

There are approaching 1000 species of sand-flies in six genera, within the subfamily Phlebotominae of the family Psychodidae. Species in three genera – *Phlebotomus*, *Lutzomyia* and *Sergentomyia* – suck blood from vertebrates; the former two are the more important as they contain disease vectors. The genus *Phlebotomus* occurs only in the Old World, from southern parts of the northern temperate areas, mainly the Mediterranean region, to central Asia, and in tropical areas, but there are not many species in sub-Saharan Africa or South-east Asia and none in the Pacific area. Most *Phlebotomus* species inhabit semiarid and savannah areas in preference to forests. *Lutzomyia* species are found only in the New World, and, by contrast, occur mainly in forested areas of Central and South America.

Sergentomyia species are also confined to the Old World, being found mainly in the Indian subregion, sub-Saharan Africa and Asia. Although a few species bite people they are not vectors.

Adult flies are often called sand-flies because of their colour. However, this can be confusing because in some parts of the world the small biting midges of the family Ceratopogonidae (Chapter 6) and black-flies (Simuliidae, Chapter 4) are called sand-flies. The medically most important species include *Phlebotomus papatasi*, *P. sergenti*, *P. argentipes*, *P. ariasi*, *P. perniciosus* and species in the *Lutzomyia longipalpis* and *L. flaviscutellata* complexes. In both the Old and New Worlds sand-flies are vectors of leishmaniasis and viruses responsible for sandfly fever, and in the Andes the bacterium *Bartonella bacilliformis*, causing bartonellosis (Carrión's disease).

5.1   External morphology

Adults of *Phlebotomus* and *Lutzomyia* are difficult to distinguish, but as the former genus is found only in the Old World and the latter in the New World this is not a problem.

Adult phlebotomine sand-flies are readily recognized by their minute size (1.5–3.5 mm in length), hairy appearance, relatively large black eyes and their long and *stilt-like* legs (Plate 6). The only other blood-sucking flies which are as small as this are some species of biting midges (Ceratopogonidae), but these have non-hairy wings and differ in many other details (Chapter 6). Phlebotomine sand-flies have the head, thorax, wings and abdomen densely covered with long hairs. The 16-segmented *antennae* are long and composed of small bead-like segments having short hairs; antennae are similar in both sexes. The mouthparts are short and inconspicuous and adapted for blood-sucking, but *only females* bite. At their base is a pair of five-segmented maxillary palps which are relatively conspicuous and droop downwards.

Wings are lanceolate in outline and quite distinct from the wings of other biting flies. The Phlebotominae can be distinguished from other subfamilies of the family Psychodidae, which they may superficially resemble, by their *wings*. In sand-flies the wings are held at an angle of about 40° over

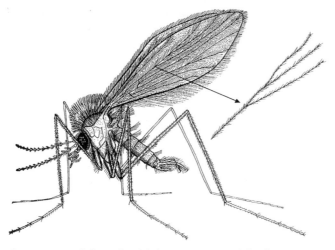

Figure 5.1 Adult male phlebotomine sand-fly showing genital claspers at end of abdomen, and a diagrammatic representation of the double branching of wing vein 2.

the body when the fly is at rest or blood-feeding, whereas in non-biting psychodid flies they are folded, roof-like, over the body or flat. Wing venation also differs. In phlebotomine sand-flies, but not in the other subfamilies of Psychodidae, *vein 2 branches twice*, although this may not be apparent unless most of the hairs are rubbed from the wing veins (Fig. 5.1).

The abdomen is moderately long and in the female more or less rounded at the tip. In males it terminates in a prominent pair of genital claspers (Fig. 5.1) which give the end of the abdomen an upturned appearance.

Identification of adult phlebotomine sand-flies to species is difficult and usually necessitates the examination of internal structures, such as the arrangement of the teeth on the cibarial armature, the shape of the spermatheca in females, and in males the structure of the external genitalia (terminalia).

5.2   Life cycle

The minute eggs (0.3–0.4 mm) are more or less ovoid in shape and usually brown or black, and careful examination under a microscope reveals that they are patterned, as shown in Figure 5.2. Some 30–70 eggs are laid singly at each oviposition. They are thought to be deposited in small cracks and holes in the ground, at the base of termite mounds, in cracks in masonry, on stable floors, in poultry houses, amongst leaf litter and in between buttress-roots of forest trees, etc. The type of oviposition site presumably varies greatly according to species.

Although eggs are not laid in water they require a microhabitat with high humidity. They are unable to withstand desiccation and hatch after 4–20

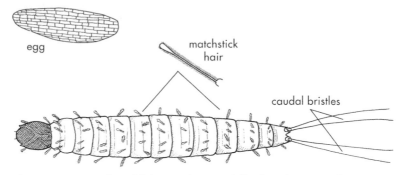

Figure 5.2 Egg of a phlebotomine sand-fly showing mosaic-type pattern, and a larva with matchstick hairs and caudal setae.

days, although hatching may likely be delayed in cooler weather. Larvae are mainly scavengers, feeding on organic matter such as fungi, decaying forest leaves, semi-rotting vegetation, animal faeces and decomposing bodies of arthropods. Although some species, especially of the genus *Phlebotomus*, occur in semiarid areas, the actual larval habitats must have a high degree of humidity. Larvae may be able to survive by migrating to drier areas if their breeding places are temporarily flooded.

There are four larval instars. The mature larva is 3–6 mm long and has a well-defined black head which is provided with a pair of small mandibles; the body is white or greyish and 12-segmented (Fig. 5.2). Ventrally the abdominal segments have small pseudopods, but the most striking feature is the presence on the head and all body segments of conspicuous thick bristles with feathered stems, which in many species have slightly enlarged tips. They are called *matchstick* hairs and identify larvae as those of phlebotomine sand-flies. In most species the last abdominal segment bears two pairs of conspicuous long hairs called the *caudal setae*. First-instar larvae have two single bristles, not two pairs.

Larval development is usually complete after 20–30 days' duration depending on species, temperature and availability of food. In temperate areas and arid regions species may overwinter as diapausing fully grown larvae. Prior to pupation the larva assumes an almost erect position in the habitat, the skin then splits open and the pupa wriggles out. The larval skin is not completely cast off but remains attached to the end of the pupa. The presence of this skin, with its characteristic two pairs of caudal bristles, aids in the recognition of the phlebotomine pupa. The pupal shape is as shown in Figure 5.3. Adults emerge from the pupae after about 6–13 days. The life cycle, from oviposition to adult emergence, is 30–60 days, but extends to several months in some species with diapausing larvae. In temperate areas adults die off in late summer or autumn and species overwinter as larvae, with the adults emerging the following spring. It is usually extremely

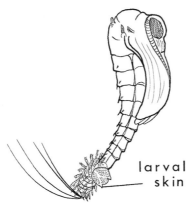

Figure 5.3 Pupa of a phlebotomine sand-fly with larval skin still attached.

difficult to find larvae or pupae of sand-flies, and relatively little is known about their biology and ecology.

5.2.1   Adult behaviour

Both sexes feed on plant juices and sugary secretions but *females* in addition suck blood from a variety of vertebrates, including livestock, dogs, urban and wild rodents, snakes, lizards and amphibians; a few species feed on birds. Females of many *Phlebotomus* species in the Old World and *Lutzomyia* species in the New World bite mammals, including humans. Biting is usually restricted to crepuscular and nocturnal periods but people may be bitten during the day in darkened rooms, or in forests during overcast days. Most species feed out of doors (*exophagic*) but some species also feed indoors (*endophagic*). A few species are autogenous, that is they can lay eggs without blood-feeding.

Adults are *weak fliers* and usually disperse 100 metres or less from their breeding places. Consequently biting is often very localized. However, adults of at least some species have been known to fly up to 2.2 km over a few days. When close to a host sand-flies may have a characteristic hopping type of flight, so that there may be several short flights and landings before females settle on a host. Even a light wind inhibits flight activities and biting. Because of their very short mouthparts sand-flies are unable to bite through clothing.

During the day adult sand-flies rest in sheltered, dark and humid sites, but on dry surfaces, such as on tree trunks, on ground litter and foliage of forests, in animal burrows, termite mounds, tree-holes, rock fissures, caves, cracks in the ground and inside human and animal habitations. Species that commonly rest in houses (*endophilic*) before or after feeding on humans are often referred to as domestic or peridomestic species. Examples are

Phlebotomus papatasi in the Mediterranean area, *P. argentipes* in India and the *Lutzomyia longipalpis* complex in South America.

In temperate areas of the Old World sand-flies are seasonal and adults occur only in the summer months. In tropical areas some species are common more or less throughout the year, but in other species there may be well-marked changes in abundance of adults related to the dry and wet seasons.

5.3   Medical importance

5.3.1   Annoyance
About 70 species are vectors of disease to humans, but apart from their importance as vectors, sand-flies may constitute a serious, but usually localized, biting nuisance. In previously sensitized people their bites may result in severe and almost intolerable irritations, a condition known in the Middle East as harara.

5.3.2   Leishmaniasis
Leishmaniasis is a term used to describe a number of closely related diseases caused by about 20 distinct species, subspecies and strains of *Leishmania* parasites in approximately 90 countries. The three main clinical forms are cutaneous, mucocutaneous and visceral leishmaniasis. A fourth, less common form is diffuse cutaneous leishmaniasis, while post-kala-azar dermal leishmaniasis is caused by *Leishmania donovani donovani* following cure of the initial visceral form. The epidemiology of leishmaniasis is complex, involving not only different parasite species but different strains of parasites and different reservoir hosts.

Basically *amastigote* parasites ingested by female sand-flies with a blood-meal multiply in the gut and develop into *promastigotes*, which are elongate, have a flagellum and attach to the mid-gut or hind-gut wall and multiply rapidly. Many, however, are voided when the fly defecates. After further development the survivors migrate to the anterior part of the mid-gut and then to the fore-gut. Here some parasites become *metacyclic* forms. Four to twelve days after the sand-fly has taken an infective blood-meal the metacyclic forms are found in the mouthparts, and are introduced into a new host during feeding. Infective flies often probe more often than uninfected flies, thus maximizing transmission of parasites during blood-feeding. Previous feeding by females on sugary substances, mostly obtained from plants, is essential not only for the survival of the sand-fly but also for the development of the parasites to the infective form.

Most types of leishmaniasis are *zoonoses*. The degree of involvement of humans varies greatly from area to area. The epidemiology is largely determined by the species of sand-flies, their ecology and behaviour, the availability of a wide range of non-human hosts, and also by the species and strains of *Leishmania* parasites. In some areas, for example, sand-flies will transmit infections almost entirely among wild or domesticated animals, with little or no human involvement, whereas elsewhere animals may be important *reservoir hosts* of infection for humans. In India the disease may be transmitted between people by sand-flies, with animals taking no identifiable part in its transmission. The epidemiology of the leishmaniases is complex, and only simplified accounts are given below.

Cutaneous leishmaniasis (CL)
In the Old World, CL is known also as *oriental sore*. It occurs mainly in arid areas of the Middle East to north-western India and central Asia, in North Africa and various areas in East, West and southern Africa. The principal parasites are *Leishmania major*, transmitted mainly by *Phlebotomus papatasi*, and *Le. tropica*, transmitted by *P. sergenti*. *Leishmania major* is usually zoonotic and in most of its range gerbils (e.g. *Rhombomys opimus*) are the reservoir hosts; *Le. tropica* occurs in densely populated areas and humans appear to be the main reservoir hosts. In the New World, CL is found mainly in forests from Mexico to northern Argentina, and is caused by *Leishmania braziliensis*, *Le. amazonensis* and *Le. mexicana*. Rodents and dogs appear to be reservoir hosts. Vectors include *Lutzomyia wellcomei* and *L. flaviscutellata*.

Mucocutaneous leishmaniasis (ML) (espundia)
A severely disfiguring disease found from Mexico to Argentina. It is mainly caused by *Leishmania braziliensis*. Dogs may be reservoir hosts. *Lutzomyia wellcomei* is an important vector.

Diffuse cutaneous leishmaniasis (DCL)
A form that causes widespread cutaneous nodules or macules over the body. It is confined to Venezuela and the Dominican Republic and the highlands of Ethiopia and Kenya. In South America the parasite is *Le. amazonensis* transmitted by *Lutzomyia flaviscutellata* and spiny rats (*Proechimys* species) are reservoir hosts. In Ethiopia and Kenya the parasite is *Le. aethiopica*, transmitted by *Phlebotomus pedifer* and *P. longipes*, with rock hyraxes (*Procavia capensis*) as reservoir hosts.

Visceral leishmaniasis (VL)
Often referred to as *kala-azar*. It is caused by *Leishmania donovani donovani* in most areas of its distribution, such as India, Bangladesh, Sudan, East Africa and Ethiopia. Among the vectors are *P. argentipes* and *P. orientalis*. Rodents, wild cats and genets (*Genetta genetta*) may be reservoir hosts. In the Mediterranean basin, Iran and central Asia, including northern and central China, *Leishmania donovani infantum* is the parasite, and the vectors include *P. ariasi* and *P. perniciosus*. Dogs and foxes (*Vulpes vulpes*) are

reservoir hosts. Visceral leishmaniasis also occurs sporadically in Central and South America, where the parasite is *Leishmania donovani infantum* (*Le. chagasi* of some authors), transmitted by species in the *Lutzomyia longipalpis* complex.

5.3.3   Bartonellosis

Bartonellosis, sometimes called Oroya fever or Carrión's disease, is encountered in arid mountainous areas of the Andes, mainly in Peru, but also in Ecuador and Colombia. It is caused by the bacterium *Bartonella bacilliformis* and is transmitted by *Lutzomyia verrucarum* and probably by other *Lutzomyia* species, such as *L. colombiana* in Colombia. Transmission is possibly only by contamination of the mouthparts.

5.3.4   Sandfly fevers

Sand-flies transmit the seven viral serotypes responsible for sandfly fevers, also called papataci fever (sometimes spelt papatasi or pappataci fever), three-day fever or *Phlebotomus* fevers. The classical form of the disease is found in the Mediterranean region, but also extends up the Nile into Egypt, and from the Middle East to northern India, Pakistan, Afghanistan and probably China. The most important vector is *P. papatasi*. Other forms of the virus in Central and South America are transmitted by *Lutzomyia* species such as *L. trapidoi*.

Female sand-flies become infective 7–10 days after taking a blood-meal. Infected females lay eggs containing the virus and these eventually give rise to infected adults. This is an example of **transovarial** transmission, a phenomenon that is more common in the transmission of various tick-borne diseases (Chapters 16 and 17). There are probably mammalian reservoir hosts.

5.4   Control

Although phlebotomine sand-flies are very susceptible to insecticides, until recently there have been few organized attempts to control them. However, in most areas where house-spraying has been used to control *Anopheles* vectors there have been large reductions in sand-fly populations followed by interruption of leishmaniasis transmission. When houses in Kabul, Afghanistan, and in the Peruvian Andes were sprayed with the pyrethroid lambda-cyhalothrin cutaneous leishmaniasis was reduced by 60% and 54% respectively. In Venezuela spraying houses with lambda-cyhalothrin substantially reduced the vector (*Lutzomyia ovallesi*) of cutaneous leishmaniasis.

Obviously where sand-flies bite and rest out of doors house-spraying will have little effect. However, if the outdoor resting sites are known (e.g. animal shelters, stone walls, tree trunks, termite hills) they can be sprayed with residual insecticides. Insecticidal fogging of outdoor resting sites may also give some, but temporary, control of vectors.

Personal protection can be achieved by applying efficient insect repellents such as DEET, piperidene-based ones and 2–5% neem oil. Mesh screens or nets with very small holes can give protection, but they reduce ventilation and cause it to be unpleasantly hot in screened houses or under sand-fly bed-nets. However, such nets, or even mosquito nets with larger holes, can be impregnated with pyrethroids (e.g. permethrin and deltamethrin) to give protection against biting for up to 6–12 months. In Afghanistan and Syria insecticide-treated polyester bed-nets gave good protection against cutaneous leishmaniasis (*Le. tropica* transmitted by *Phlebotomus sergenti*).

In some countries there is a combined campaign of insecticidal spraying of houses and the distribution of insecticide-treated bed-nets, a strategy which targets both *Anopheles* mosquitoes and sand-flies (see control measures, Chapter 2).

Control of sand-fly larvae remains impossible because the breeding sites of most species are unknown.

Because most leishmaniasis transmission involves ***reservoir hosts***, such as rodents and dogs, attempts have been made to destroy them. In China leishmaniasis was effectively eliminated in the 1950s by killing dogs, but although similar culls have been made in parts of Brazil and the Mediterranean region results have been disappointing. Dogs have sometimes been dipped or sprayed with pyrethroids such as deltamethrin, but repeated treatments, typically every 2–3 months, are needed. Deltamethrin-treated collars on dogs, which can remain effective for 8 months, have also given good, albeit local, control of *Le. donovani infantum* in Italy and Iran. Trials in the Mediterranean with a new vaccine against visceral leishmaniasis have proved promising in protecting dogs, thus reducing the numbers of reservoir hosts. In Russia and Jordan zoonotic cutaneous leishmaniasis has been controlled by destroying rodent colonies, but elsewhere results have not been encouraging.

Insecticide resistance to DDT has been found in *P. papatasi* and *P. argentipes* in India, and greater tolerance or resistance to several insecticides, such as pyrethroids, has been reported in other sand-flies.

In 2005 India, Bangladesh and Nepal signed an agreement to eliminate visceral leishmaniasis from the Indian subcontinent, mainly by integrated vector control.

Further reading

Alexander, B. and Maroli, M. (2003) Control of phlebotomine sandflies. *Medical and Veterinary Entomology*, **17**, 1–18.

Ashford, R. W. (2001) Leishmaniasis. In *The Encyclopedia of Arthropod-Transmitted Infections of Man and Domesticated Animals*, ed. M. W. Service. Wallingford: CABI, pp. 269–79.

Guerin, P. J., Olliaro, P., Sundar, S. *et al.* (2002) Visceral leishmaniasis: current status of control, diagnosis and treatment, and a proposed research and development agenda. *Lancet Infectious Diseases*, **2**, 494–501.

Hide, G., Mottram, J. C., Coombs, G. H. and Holmes, P. H. (1996) *Trypanosomiasis and Leishmaniasis: Biology and Control*. Wallingford: CAB International.

Killick-Kendrick, R. (1990) Phlebotomine vectors of the leishmaniases: a review. *Medical and Veterinary Entomolology*, **4**, 1–24.

Killick-Kendrick, R. (1999) The biology of phlebotomine sand flies. *Clinics in Dermatology*, **17**, 279–89.

Lainson, R. (1983) The American leishmaniases: some observations on their ecology and epidemiology. *Transactions of the Royal Society of Tropical Medicine and Hygiene*, **77**, 569–96.

Lainson, R. (1989) Demographic changes and their influence on the epidemiology of the American leishmaniases. In *Demography and Vector-Borne Diseases*, ed. M. W. Service. Boca Raton, FL: CRC Press, pp. 85–106.

Lane, R. P. (1991) The contribution of sand-fly control to leishmaniasis control. *Annales de la Société Belge de Médicine Tropicale*, **71** (suppl.), 65–74.

Peters, W. and Killick-Kendrick, R. (eds.) (1987) *The Leishmaniases in Biology and Medicine. Volume 1: Biology and Epidemiology. Volume 2: Clinical Aspects and Control*. London: Academic Press.

Tayeh, A., Jalouk, L. and Al-Khiami, A. M. (1997) Cutaneous leishmaniasis control trial using pyrethroid-impregnated bednets in villages near Aleppo, Syria. WHO/LEISH/**97.41**. *Geneva: World Health Organization, Division of Control of Tropical Diseases.*

Ward, R. D. (1990) Some aspects of the biology of phlebotomine sand-fly vectors. *Advances in Disease Vector Research*, **6**, 91–126. (Reprints and chapters incorrectly dated 1989.)

World Health Organization (1990) Control of the leishmaniases: report of a WHO Expert Committee. *World Health Organization Technical Report Series*, **793**, 1–158.

6

Biting midges (Ceratopogonidae)

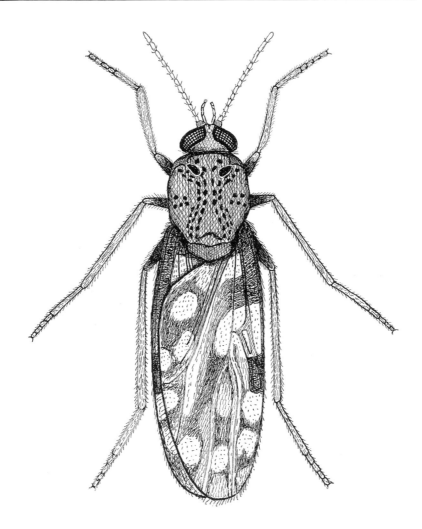

There are about 5540 species of biting midges in some 103 genera, but only four genera feed on vertebrates. Medically the most important two genera are *Leptoconops*, which is mainly found in the tropics and subtropics, including the Caribbean area and parts of the USA, and *Culicoides*, which contains about 1450 species and has an almost worldwide distribution. In many parts of the world species of *Culicoides*, and in the Americas also *Leptoconops*, can constitute serious biting problems. In Africa *Culicoides milnei*, *C. austeni* and possibly *C. grahamii* are vectors of the filarial worms *Mansonella perstans*, while *C. grahamii* and possibly *C. milnei* and *C. austeni* are vectors of *Mansonella streptocerca*. *Culicoides furens* is a vector of *Mansonella ozzardi* in the Americas. These parasites are usually regarded as non-pathogenic to humans. Although they can be a serious biting nuisance, *Leptoconops* are not disease vectors.

6.1   External morphology

Adults are sometimes known as midges or biting midges, and, especially in the Americas, as 'no-see-ums'. In Australia and some other countries they are often called sand-flies, but this name is unfortunate and should be avoided because phlebotomines (Chapter 5) and occasionally simuliids (Chapter 4) may also be referred to as sand-flies. The most appropriate common name is biting midges; this terminology serves to distinguish them from other small non-biting flies which are often referred to as midges.

Adult *Culicoides* (Plate 7) are very small, being only 1–2.5 mm long, and with the phlebotomines constitute the smallest biting flies attacking humans.

The head has a prominent pair of eyes, a pair of short five-segmented palps and a pair of relatively long filamentous antennae. As in mosquitoes, males do not take blood-meals and have feathery or **plumose antennae**, whereas the **blood-sucking females** have **non-plumose** antennae. The biting mouthparts, which are very small and inconspicuous, do not project forwards but hang down from the head. The arrangement and structure of the mouthparts are very similar to those of simuliids (Chapter 4). In some species the thorax is covered dorsally with distinct very small black spots and other markings. In all *Culicoides* species there is dorsally on the anterior part of the thorax a pair of small elongate shiny black depressions, known as **humeral pits**. Their presence distinguishes the genus *Culicoides* from *Leptoconops*. The wings are short and relatively broad, and apart from the first veins venation is faint. Wings lack scales but in many species are covered with minute hairs, but these are seen only under a dissecting microscope. In most *Culicoides* species the density of these hairs gives the **wings** the appearance of having contrasting dark and milky white spots or patches (Fig. 6.1). In life the wings are placed over the abdomen like the blades of a closed pair of scissors (Fig. 6.1). The legs are relatively short.

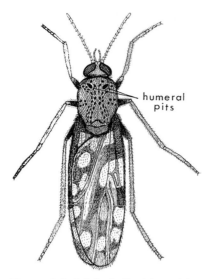

Figure 6.1 Adult *Culicoides* midge, showing patterned wings and the two thoracic humeral pits.

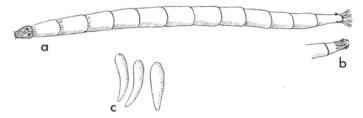

Figure 6.2 *Culicoides*: (a) larva with two four-lobed retractile papillae extending beyond the last abdominal segment; (b) last abdominal segment with papillae withdrawn into it; (c) eggs.

The abdomen is dull grey, yellowish brown or blackish, and in females is more or less rounded at its tip; in males there is a pair of small, but conspicuous, claspers.

6.2   Life cycle

Eggs are dark, cylindrical or curved and banana-shaped, and about 250–500 μm long (Fig. 6.2c). They are laid in batches of about 30–350 on the surface of mud, wet soil, especially that near swamps and marshes including saltwater marshes, on decaying leaf litter, humus, manure, or on plants and other objects near or partially submerged in water. They are also laid in tree-holes, in semi-rotting vegetation and in the cut stumps of banana plants (for example *Culicoides milnei*, *C. austeni* and *C. grahamii*). The type of oviposition site selected depends on the species.

Eggs usually hatch after 2–9 days, depending on temperature and species; some temperate species overwinter as eggs. There are four larval instars and a fully grown larva is cylindrical, whitish, about 3–6 mm long and nematode-like. It has a small, dark (but unpigmented in *Leptoconops*) conical-shaped head followed by 12 body segments. The last segment terminates in a pair of four-lobed *retractile papillae* (Fig. 6.2a). These are not always readily seen in preserved larvae because they are often retracted within the last abdominal segment (Fig. 6.2b). *Culicoides* larvae are best recognized by the combination of a small dark head followed by a segmented body devoid of any obvious structures and, when they are extruded, by the presence of terminal papillae. When alive they can also be recognized by their serpentine swimming motions.

Larvae feed mainly on decaying vegetable matter. When the water level in swamps and marshes rises larvae of many species migrate to the damp soil and mud at the edges to avoid drowning. Some important pest species breed in sandy areas near the seashore. Larvae are difficult to find and are rarely encountered unless special surveys are made to collect them.

Larval development is sometimes completed within 4–5 days, but may take weeks in cooler weather, and in temperate regions many species overwinter as larvae for seven months or more. Species occupying marshy habitats frequently migrate to the drier peripheral areas for pupation. However, in species that are aquatic the pupae float at the water surface.

The pupa (Fig. 6.3) is 2–4 mm long and readily recognized by the following combination of characters: (1) a pair of breathing trumpets on the cephalothorax which appear to be composed of two segments; (2) abdominal segments bearing small but conspicuous tubercles ending in a fine hair; and (3) a prominent pair of horn-like processes on the last abdominal segment. The pupal period usually lasts 2–3 days.

6.2.1 Adult behaviour

Adults of both sexes feed on naturally occurring sugar solutions. In addition *females* take blood-meals from humans and a variety of mammals and birds. Adults bite at any time of the day or night, but many species are particularly active and troublesome in the evenings. In contrast *C. grahamii*, an African species, bites mainly in the early part of the mornings. Because of their short mouthparts biting midges, unlike mosquitoes, tabanids and tsetse-flies, can rarely bite through clothing. For this reason midges often swarm around the head, biting the face, especially the forehead and scalp. They also bite the hands and exposed arms and legs. Most species bite only outdoors, but a few including the African species *C. milnei*, *C. austeni* and *C. grahamii* will enter houses to feed on people (*endophagic*). About 30 species are *autogenous*, that is they can mature and lay the first batch of eggs without a blood-meal. Adults usually fly only a few hundred metres from their larval habitats, but sometimes they are wind-dispersed

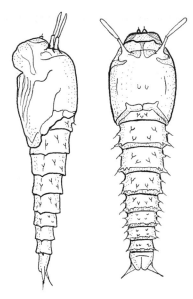

Figure 6.3 *Culicoides* pupa, lateral and dorsal views. Note the two long respiratory trumpets.

considerably further. However, some species of both *Culicoides* and *Leptoconops* are known to fly 2–3 km without wind assistance. Adult females are short-lived, most probably surviving for just 1–3 weeks.

6.3   Medical importance

The Ceratopogonidae are not very important vectors of disease to humans, whereas they are of considerable veterinary importance, transmitting arboviruses such as those causing bluetongue disease in sheep and African horse sickness. They can, however, be very troublesome pests and transmit minor filarial infections as well as Oropouche virus to humans (see below).

6.3.1   Annoyance

Biting midges are very small, but what they lack in size they can make up for in numbers – as has been said, one midge is an entomological curiosity, a thousand sheer hell! In areas as dissimilar as the west coast of Scotland, the Caribbean, California and Florida, biting midges can be a serious economic threat to the tourist industry. Persistent biting of large numbers of midges can make outdoor recreational activities impossible, not only at dusk but often during much of the day. In some areas they have prevented the continuation of harvesting and other outdoor work during the evenings.

An important pest species in southern areas of North America down to Brazil is *Culicoides furens*, which breeds in salt marshes and other saline coastal habitats. In North America *Leptoconops torrens* and *L. bequaerti*,

which breed in sandy soils and coastal areas, can be serious pests. In Europe
C. impunctatus and many other species are troublesome biters, while in
Madagascar, the Seychelles and Brunei *L. spinosifrons* can cause a considerable biting problem.

6.3.2   Filarial infections
A few *Culicoides* species are vectors of filarial parasites to humans.

Mansonella perstans
Found in Africa, especially West and Central, but also parts of East Africa
as far south as Zimbabwe. Vectors include *C. milnei* and *C. austeni*, and
possibly also *C. grahamii*. These species breed in the rotting cut stumps of
banana and plantain plants. *Mansonella perstans* also occurs in Central and
South America and Trinidad, where it is transmitted by other *Culicoides*
species.

Mansonella streptocerca
Occurs in the rainforests of West Africa, from Côte d'Ivoire to Gabon, the
Republic of the Congo, the Democratic Republic of Congo (formerly Zaire)
and also western Uganda, where transmission is by *C. grahamii* and possibly
by *C. milnei* and *C. austeni*. These species breed in the rotting cut stumps of
banana and plantain plants.

Mansonella ozzardi
Found in Mexico, Panama, the Caribbean islands and South America.
Transmission is by *Culicoides* species, mainly by *C. furens* and *C. phlebotomus*. (The role of simuliids in the transmission of this filarial parasite is
discussed in Chapter 4.)

Microfilariae of all these parasites are **non-periodic** and are ingested during feeding. They undergo morphological changes, invade the thoracic
flight muscles, moult twice and then migrate to the head and after about
8–12 days pass down the proboscis. Infective **third-stage** larvae (L_3) are
deposited on the skin of the host when the female takes a blood-meal. The
salivary glands of *Culicoides* play no part in the transmission of these parasites. None of these three filarial parasites appears to cause much harm
and they are usually regarded as non-pathogenic, although morbidity or
allergic reactions may sometimes occur.

6.3.3   Arboviruses
The only known arbovirus biologically transmitted to humans by midges
is Oropouche virus (*Bunyavirus*) which occurs in Brazil, Panama, Peru,
Trinidad and possibly Colombia. Transmission is by the bite of *Culicoides
paraensis*.

6.4   Control

It is usually very difficult to control the larvae because many larval habitats are extensive, such as freshwater and saltwater marshes and wet coastal sands, whose limits may be difficult to identify.

Although larval habitats could be eradicated by draining them or filling them in, this is usually too costly and laborious, and in many situations impractical. However, if undertaken and maintained, such methods have the advantage of giving permanent control. Although such environmental control avoids contaminating the environment with insecticides it nevertheless results in the loss of ecological habitats.

Spraying larval breeding sites with malathion, diazinon or temephos has sometimes been effective, although heavy rainfall is often required to wash the insecticide through surface vegetation to the underlying soil and mud harbouring the larvae. Spraying may require repeating after 1–2 months, and increased public awareness of environmental contamination might prevent this approach.

A few species enter buildings to bite but bed-nets and screening used to exclude house-flies and mosquitoes will not keep out the much smaller midges, unless they are impregnated with pyrethroids such as permethrin and deltamethrin. Such treated nets may remain effective for six months, or for five years if long-lasting nets are used. (For more details on the use of insecticide-impregnated bed-nets against vectors see Chapter 2.) Alternatively protective nets and screens having very small mesh size can be used, but they will substantially reduce air flow and ventilation.

Insecticidal fogs or ultra-low-volume (ULV) applications have sometimes been used to kill adults resting in vegetation, but the effects are very short-lived and sprayed areas are soon invaded by midges flying in from unsprayed areas.

Limited personal protection can be achieved by using repellents such as DEET.

Further reading

Borkent, A. (2005) The biting midges, the Ceratopogonidae (Diptera). In *Biology of Disease Vectors*, 2nd edn, ed. W. C. Marquart. Amsterdam: Elsevier Academic Press, pp. 113–26.

Halouzka, J. and Hubalek, Z. (1996) Biting midges (Ceratopogonidae) of medical and veterinary importance: a review. *Acta Scientifiarum Naturalium Academiae Scientiarum Bohemica, Brno* **30** (2), 1–56.

Kettle, D. S. (1965) Biting ceratopogonids as vectors of human and animal diseases. *Acta Tropica*, **22**, 356–62.

Kettle, D. S. (1969) The ecology and control of blood-sucking ceratopogonids. *Acta Tropica*, **26**, 235–48.

Kettle, D. S. (1977) Biology and bionomics of blood-sucking ceratopogonids. *Annual Review of Entomology*, **22**, 33–51.

Linley, J. R. and Davies, J. B. (1971) Sandflies and tourism in Florida and the Bahamas and Caribbean area. *Journal of Economic Entomology*, **64**, 264–78.

Linley, J. R., Hoch, A. L. and Pinheiro, F. P. (1983) Biting midges (Diptera: Ceratopogonidae) and human health. *Journal of Medical Entomology*, **20**, 347–64.

Mellor, P. S. (2001) Oropouche virus. In *The Encyclopedia of Arthropod-transmitted Infections of Man and Domesticated Animals*, ed. M. W. Service. Wallingford: CABI, pp. 391–3.

Mellor, P. S., Boorman, J. and Baylis, M. (2000) *Culicoides* biting midges: their role as arbovirus vectors. *Annual Review of Entomology*, **45**, 307–40.

Nathan, M. B. (1979) The prevalence and distribution of *Mansonella ozzardi* in coastal north Trinidad, W. I. *Transactions of the Royal Society of Tropical Medicine and Hygiene*, **73**, 299–302.

7

Horse-flies (Tabanidae)

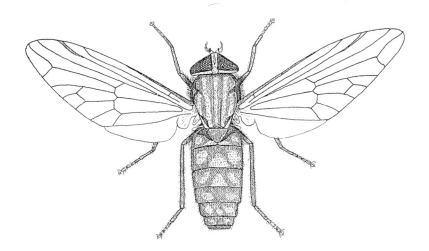

Tabanids are large biting flies generally called horse-flies, although other vernacular names include greenheads (some species of *Tabanus*), clegs and stouts (*Haematopota*) and deer-flies (*Chrysops*). All belong to the family Tabanidae, which comprises some 4300 species and subspecies in 133 genera. Medically the most important are species of *Tabanus*, *Chrysops* and *Haematopota*. Tabanids have been incriminated in the spread of anthrax and tularaemia and might be involved in the transmission of Lyme disease (usually transmitted by hard ticks). But their main medical importance is that species of *Chrysops*, mainly *C. silaceus* and *C. dimidiatus*, are vectors in West and Central Africa of the filarial worm *Loa loa*.

The Tabanidae have a worldwide distribution. Species of *Tabanus* and *Chrysops* are found in temperate and tropical areas, but *Haematopota* is absent from South America and Australasia and is uncommon in North America.

7.1   External morphology

A generalized description is presented of the Tabanidae, with special reference to the genera *Chrysops*, *Tabanus* and *Haematopota*.

Tabanids are medium to very large flies (6–30 mm long). Many, especially *Tabanus* species, are robust and heavily built, and this genus contains the largest biting flies, some with a wingspan of 65 mm. The colouration of tabanids varies from very dark brown or black to lighter reddish brown, yellow or greenish; frequently the abdomen and thorax have stripes or patches of contrasting colours (Fig. 7.1). The head is large and, viewed from above, is more or less *semicircular* (Fig. 7.2); it is often described as semilunar. The head has a conspicuous pair of compound eyes which in life may be marked with contrasting *iridescent* colours, such as greens and reds or even purplish hues, arranged in bands, zigzags or spots. Adults are *sexed* by examining their eyes. In the female there is a distinct space on top of the head separating the eyes: this is known as a *dichoptic* condition (Fig. 7.2a). In females of some species this space between the eyes may be narrow, whereas in others, especially *Chrysops*, it is quite large. In males the eyes are so large that they occupy almost all of the head and either touch each other on top of the head or are very narrowly separated, this being known as a *holoptic* condition (Fig. 7.2b).

The *antennae* are relatively small but stout. They consist of three segments; the last is subdivided into usually three or four small divisions by annulations. Unlike the Muscidae, Glossinidae and Calliphoridae, there is no antennal arista. The size and shape of the antennae serve to distinguish the genera *Chrysops*, *Haematopota* and *Tabanus* (Fig. 7.3). The mouthparts of female Tabanidae are stout and adapted for biting and, unlike those of tsetse-flies, mosquitoes and *Stomoxys*, they do not project forwards but hang down from the head. Only *female* tabanids take blood-meals.

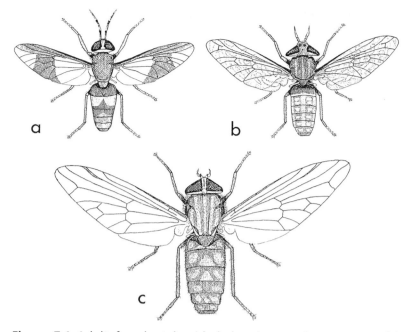

Figure 7.1 Adult female tabanids belonging to three genera: (a) *Chrysops*; (b) *Haematopota*; (c) *Tabanus*.

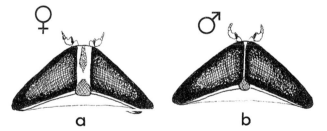

Figure 7.2 Dorsal view of tabanid heads: (a) female showing dichoptic eyes; (b) male with holoptic eyes.

The stout thorax bears a pair of wings which have two submarginal and five posterior cells and a completely closed *discal cell* in approximately the centre of the wing (Fig. 7.4). Although wing venation alone may not be sufficient to identify the Tabanidae from all other Diptera, it nevertheless serves as a useful guide when considered with other characters, such as the shape and structure of the antennae and biting mouthparts. Wings may be completely clear and devoid of colour, have areas of brown colouration, be distinctly banded, or appear mottled or speckled due to greyish patches (Fig. 7.1).

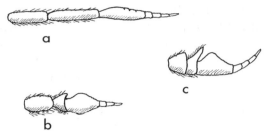

Figure 7.3 Antennae of adult tabanids belonging to three genera: (a) *Chrysops*; (b) *Haematopota*; (c) *Tabanus*.

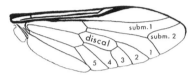

Figure 7.4 Wing of an adult tabanid showing discal cell, two sub-marginal cells (subm.) and five posterior cells (1–5).

Adults at rest have the wings placed either like a pair of open scissors over the abdomen or at a roof-like angle completely obscuring the abdomen. The presence or absence of coloured areas on the wings and the way in which they are held over the body provides useful additional characters for distinguishing between *Chrysops*, *Tabanus* and *Haematopota* (see below).

The abdomen is usually broad and stout, and in unfed flies characteristically flattened dorsoventrally. It may be a more or less uniformly dark brown, blackish, light brown, reddish brown, yellowish or even greenish, or alternatively marked with contrasting coloured stripes or patches.

7.1.1   Identification of adult *Chrysops*, *Tabanus* and *Haematopota*
Chrysops species (deer-flies)
Chrysops species (Plate 8) are medium-sized flies about 6–12 mm in size. In life most species have *iridescent* eyes, commonly with spots of red, green or purple. The wings are held partially over the abdomen in an open scissor-like fashion, and usually have one or more brownish transverse bands (Fig. 7.1a). In many species the abdomen is blackish with orange or yellow patches or bands.

The most reliable method of distinguishing *Chrysops* from *Tabanus* and *Haematopota* is by their *antennae*. In *Chrysops* the second segment of the antenna is long and cylindrical and lacks any projection (Fig. 7.3a), while the third segment has four small subdivisions. The hind tibiae have apical spurs which are absent in *Tabanus* and *Haematopota*.

Chrysops has a worldwide distribution.

Tabanus species (horse-flies, greenheads)

Tabanus species are medium to very large flies. Their *eyes* are frequently brownish but may be iridescent, with markings usually in the form of horizontal bands. Their wings, which are held over the body much as in *Chrysops*, are often clear (Fig. 7.1c), but in some species they have dark markings.

Tabanus species are readily identified by the shape and size of the *antennae*. Both the second and third antennal segments have small but distinct projections dorsally (Fig. 7.3c), while the third segment has four small subdivisions and is usually distinctly curved upwards. The antennae are much shorter than those of *Chrysops* species, and are therefore less conspicuous.

Tabanus has a worldwide distribution.

Haematopota species (clegs, stouts)

Haematopota species are medium-sized dark grey flies which are easily distinguished from *Tabanus* and *Chrysops* because in life the wings are folded *roof-like* over the abdomen. Moreover, in nearly all species the wings are *dusty grey* and speckled or mottled (Fig. 7.1b). The *eyes* have zigzag bands of iridescent colours. Antennae are rather similar to those of *Tabanus* but are usually a little longer. The third segment is straight, not curved as in *Tabanus*, has only three, not four, small subdivisions and does not bear a dorsal projection (Fig. 7.3b).

Haematopota species are not found in South America or Australasia, and only five species occur in North America. They are common in Europe, Asia, Africa and the Far East.

7.2   Life cycle

Both males and females feed on naturally occurring sugary secretions, but in addition *females* bite a wide variety of mammals including humans, domesticated animals, especially horses and cattle, deer and many other herbivores, as well as carnivores and monkeys. A few species attack birds.

Oviposition sites overhang, or are adjacent to, the larval habitats, which are often muddy, aquatic or semi-aquatic (p. 116). Some 100–1000 eggs, the number depending on the species, are firmly glued in an upright position in a single large mass (up to 25 mm long) on the underside of objects such as leaves, grassy vegetation, twigs, small branches, stones and rocks. They are covered with a secretion that makes them waterproof and are often arranged in a more or less lozenge-shaped pattern (Fig. 7.5a). Eggs are mostly creamy white, greyish or blackish, 1–3 mm long, and curved or approximately cigar-shaped. They hatch after 4–14 days, the time depending on both temperature and species. After wriggling out of the eggs the young larvae drop down on to the underlying mud or water.

Larvae are cylindrical and rather pointed at both ends (Fig. 7.5c). They are creamy white, brown or even greenish but often have darkish

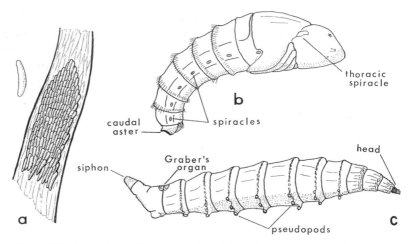

Figure 7.5 Immature stages of tabanids: (a) single egg and egg mass glued to a piece of grass; (b) pupa; (c) larva.

pigmentation near the borders of the segments. The very small black head can be retracted into the thorax. There are 11–12 well-differentiated body segments. Larvae are readily recognized by the prominent raised *tyre-like rings* which encircle most body segments. The first seven abdominal segments have one pair of lateral and two pairs of ventral (a total of six) conspicuous roundish protuberances called *pseudopods*. The presence of prominent rings and these pseudopods readily identify larvae of tabanids. The last abdominal segment has dorsally a short *siphon* which can be retracted into the abdomen

Larvae live in mud, rotting vegetation, damp soil, in shallow and often muddy waters, at the edges of small pools, swamps, ditches or slowly flowing streams. In aquatic habitats larvae sometimes adhere to floating leaves, logs or other debris. A few species breed in tree-holes and others in brackish habitats. Larvae of certain *Tabanus* and *Haematopota* species, however, occur in drier soils of pastures and in the earth near the bases of trees. Larvae breathe through their abdominal short siphon, and because they are poor swimmers aquatic species are usually found in only shallow waters. They move rather sluggishly in their muddy, aquatic or semi-aquatic habitats. Larvae of *Tabanus* and *Haematopota* are predaceous or cannibalistic. Carnivorous larvae inject a venom into their prey and occasionally cause considerable pain by biting people working barefoot in ricefields. In some species, in particular *Chrysops*, larvae are mainly scavengers, feeding on detritus and a variety of dead and decaying vegetable and animal matter.

Larval development is of a *long duration*. In both temperate and tropical countries many species remain as larvae for 1–2 years, and several cold-temperate species have a larval period of up to three years. Relatively little is known about their life cycle, but there appear to be 6–13 larval instars. Depending upon the species mature larvae may be 10–60 mm long.

Prior to pupation mature larvae migrate to drier areas at the periphery of larval habitats, where they pupate. The pupa is partially buried in the mud or soil in an upright position and, superficially, looks like a chrysalis of a butterfly (Fig. 7.5b). It is 6–35 mm long, size depending on the species, distinctly curved and usually brown. The head and thorax are combined to form a distinct cephalothorax which has a pair of lateral and relatively large ear-shaped spiracles. The abdomen has eight well-defined segments; the first seven have a pair of lateral spiracles, and segments 2–7 have an encircling row of small backwardly directed spines. The short terminal eighth abdominal segment is provided with six lobes which bear spine-like processes, known collectively as the *caudal aster*. The pupal period lasts about 5–20 days.

7.2.1   Adult behaviour

Females of most species feed during the *daytime* and are especially active in bright sunshine, although a few species are crepuscular and some feed at night. They locate their hosts mainly by sight, although olfactory stimuli such as carbon dioxide and other host odours also play a role in host location. Tabanids are powerful fliers and may disperse several kilometres.

Most tabanids inhabit woods and forests. Many *Chrysops* species are common in low-lying marshy scrub areas or swampy woods; some species, however, are found in more open savannah and grassland areas. Adults do not usually enter houses to feed, but *C. silaceus* in Africa is an exception. It is also attracted by smoke from wood burning and from forest fires.

Because of their large and blade-like mouthparts, bites are painful and wounds inflicted by tabanids frequently continue to bleed after the female has departed. Because of their painful bites tabanids are frequently disturbed when feeding, and several small blood-meals may be taken from the same or different hosts before the female has obtained a complete meal. Such interrupted feeding increases their likelihood of being mechanical vectors of disease. Because of their attraction to dark objects they often bite through coloured clothing when attacking Caucasians rather than biting exposed areas of pale skin; in this respect they behave like tsetse-flies.

In both temperate and tropical areas the occurrence of adults is seasonal. In temperate countries adults usually die off at the end of the summer and a new population emerges in the following spring or summer. In the tropics tabanids may not completely disappear in the dry months, but their numbers are normally much reduced. Maximum numbers of biting flies usually appear towards the beginning of the rainy season.

7.3   Medical importance

7.3.1   Minor infections

Because of their painful bites, tabanids may sometimes be troublesome pests and can make outdoor activities, whether recreational or work, difficult. People can develop severe allergic symptoms due to the large amount of saliva that is pumped into bite wounds to prevent blood-clotting.

Because females tend to be intermittent feeders, and are often disturbed during feeding, tabanids are particularly liable to be mechanical vectors, and in this way can spread anthrax (*Bacillus anthracis*). In North America and Russia they have been incriminated in the mechanical spread of tularaemia (*Francisella tularensis*) from horses, rabbits and rodents to humans. *Chrysops discalis* has been identified as a vector of tularaemia in North America, but other tabanid species are most likely involved. The disease is also commonly spread by handling infected rodents, by ixodid tick bites, and by eating insufficiently cooked meat. Lyme disease (*Borrelia burgdorferi*), which is normally transmitted by ixodid ticks, may also be transmitted mechanically by tabanids. They can also mechanically transmit *Trypanosoma vivax*, causing trypanosomiasis in cattle in Africa and Latin America.

Tabanids transmit viruses, bacteria, protozoa and filarial worms to livestock and therefore are of veterinary importance.

7.3.2   Loiasis

The only important and cyclically transmitted disease spread to humans by tabanids is loiasis, caused by the nematode *Loa loa* which undergoes a developmental cycle in the fly. This disease occurs principally in the equatorial rainforests of Ghana across to Nigeria and Cameroon, Equatorial Guinea, Gabon, Republic of Congo, Democratic Republic of Congo (formerly Zaire), northern Angola, southern Sudan and into western parts of Uganda. The **diurnally periodic** microfilariae are more or less absent from peripheral blood of people at night but appear in it during the day, especially in the morning. The microfilariae are therefore readily picked up by *Chrysops silaceus* and *C. dimidiatus*, species which bite during the day. In areas of Africa such as in Bahr-el-Ghazal, Sudan, where these flies are absent, other species, such as *C. distinctipennis* and *C. longicornis*, appear to be vectors.

After an infective blood-meal many, but not all, ingested microfilariae survive the process of blood digestion, penetrate the gut wall and migrate to the abdomen, and to a lesser extent the thorax and head. Here they moult twice and develop into **third-stage** larvae (L$_3$) (2 mm long), which migrate to the thorax and head, and after 7–15 days congregate in the proboscis. When *Chrysops* feed on humans as many as 200 L$_3$ larvae may be deposited on the skin. Most die, but some manage to pass through the punctures made by the biting flies, or through skin abrasions, and enter the host. They migrate to the connective tissues and become mature worms in three months, and after a further three months microfilariae appear in the peripheral blood.

Loa loa has also been found in monkeys. The microfilariae appear in their peripheral blood at night and are picked up by *C. centurionis* and *C. langi*, species which are mainly crepuscular and nocturnal and which feed in the tree canopy, where monkeys are sleeping. However, laboratory experiments suggest that transmission of *L. loa* from monkeys to humans is unlikely.

7.4   Control

There are very few practical measures to control tabanids. Efficient drainage of larval habitats might reduce adult production, but problems of locating larval habitats and the cost of drainage usually prevent this approach. Because of the difficulty of locating, often extensive, breeding sites, larviciding is often logistically impossible. Moreover, because larvae of many species live in the ground large dosage rates would be needed for any insecticide to penetrate through surface soil and vegetation to reach the larvae; these problems are somewhat similar to those encountered in the control of ceratopogonid larvae (Chapter 6).

Some level of local control can sometimes be achieved by employing attractant traps such as coloured screens coated with adhesives to catch adult flies. But there are no really effective control methods against tabanids. Some degree of personal protection may, however, be obtained by using insect repellents.

Further reading

Anderson, J. F. (1985) The control of horse flies and deer flies (Diptera: Tabanidae). *Myia*, **3**, 547–98.

Anthony, D. W. (1962) Tabanids as disease vectors. In *Biological Transmission of Disease Agents*, ed. K. Maramorosch. Symposium held under the auspices of the Entomological Society of America, Atlantic City, 1960. New York, NY: Academic Press, pp. 93–107.

Cheke, R. A., Mas, J. and Chainey, J. E. (2003). Potential vectors of loiasis and other tabanids on the island of Bioko, Equatorial Guinea. *Medical and Veterinary Entomology*, **17**, 221–3.

Chippaux, J.-P., Bouchité, B., Demanov, M., Morlais, I. and LeGoff, G. (2000) Density and dispersal of the loiasis vector *Chrysops dimidiata* in southern Cameroon. *Medical and Veterinary Entomology*, **14**, 339–44.

Duke, B. O. L. (1972) Behavioural aspects of the life cycle of *Loa*. In *Behavioural Aspects of Parasite Transmission*, ed. E. K. Canning and C. A. Wright. London: Academic Press, pp. 97–107.

Foil, L. D. (1989) Tabanids as vectors of disease agents. *Parasitology Today*, **5**, 88–95.

Noireau, F., Nzoulani, A., Sinda, D. and Itoua, A. (1990) Transmission indices of *Loa loa* in the Chaillu Mountains, Congo. *American Journal of Tropical Medicine and Hygiene*, **43**, 282–8.

Thomson, M. C., Obsomer, V., Kamgno, J. *et al.* (2004) Mapping the distribution of *Loa loa* in Cameroon in support of the African Programme for Onchocerciasis Control. *Filaria Journal*, **3**, 7. www.filariajournal.com.

8

Tsetse-flies (Glossinidae)

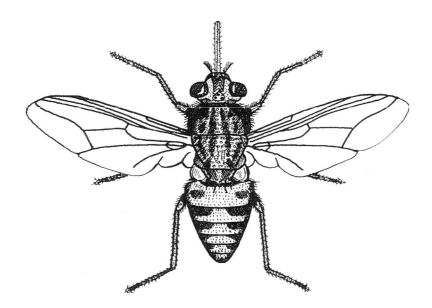

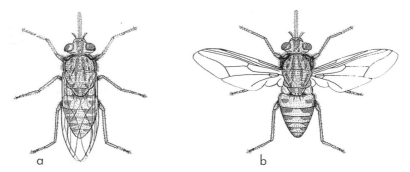

a   b

Figure 8.1 Tsetse-flies: (a) wings folded over the body like a pair of closed scissors; (b) wings purposely spread out to display abdomen and wing venation.

There are 23 species of tsetse-flies, six of which are divided into two or more subspecies all belonging to the genus *Glossina*, the only genus in the family Glossinidae. Apart from two species found in south-west Arabia, tsetse-flies are restricted to sub-Saharan Africa from approximately latitude 10° north to 20° south, but extending to 30° south along the eastern coastal area. Some species, such as *Glossina morsitans*, are found across West Africa to Central and East Africa, whereas others are more restricted in their distribution. For example, *G. palpalis* occurs only in the West African subregion.

Tsetse-flies are vectors of both human and animal African trypanosomiasis, the disease in humans being called sleeping sickness. The most important vectors are *G. palpalis*, *G. tachinoides*, *G. fuscipes*, *G. pallidipes* and *G. morsitans*.

8.1   External morphology

A general description of tsetse-flies, without special reference to any particular species, is as follows. Adults are yellowish or brown-black robust flies that are rather larger (6–14 mm) than house-flies. Some species have the abdominal segments uniformly coloured, whereas in others they may have lighter-coloured transverse stripes and a median longitudinal one. Tsetse-flies are distinguished from other flies by the combination of (1) a rigid forward-projecting *proboscis* and (2) a closed cell between wing veins 4 and 5 which, with a little imagination, looks like an upside-down hatchet (i.e. axe, cleaver or chopper) and consequently is often termed the *hatchet cell* (Figs. 8.1b, 8.2a, Plate 9). This latter character serves to conclusively identify a tsetse-fly. At rest tsetse-flies also differ from most flies by having the *wings* placed over the abdomen like the closed blades of a pair of scissors (Fig. 8.1a).

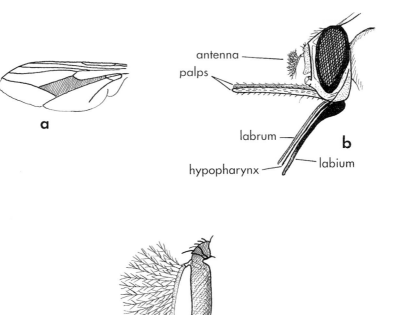

Figure 8.2 Adult tsetse-fly: (a) wing showing the hatchet cell identified by shading; (b) lateral view of head showing mouthparts, antenna and palps; (c) antenna showing plumose branching of hairs on upper side of the arista.

The proboscis is relatively large and has a bulbous base. When a tsetse-fly feeds saliva containing anticoagulants is pumped down into the wound formed by the fly. A long pair of palps occur dorsally, very close to the proboscis, and lie alongside it. They are difficult to distinguish except when the tsetse-fly is feeding and the proboscis is swung downwards while the palps remain projecting forwards (Fig. 8.2b). The first two antennal segments are small and inconspicuous but the third is relatively large, cylindrical and somewhat banana-shaped. Near its base is the arista, which has branched hairs, but only on the upper surface (Fig. 8.2c).

Dorsally the thorax has dark brown stripes or patches. There are eight abdominal segments, which may be uniformly dark brown or blackish, or have pale brown or yellowish transverse stripes. Because *both sexes* take blood-meals and can be vectors it is not important to distinguish them, nevertheless tsetse-flies can be sexed by examining the tip of the abdomen. In the males there is ventrally a prominent raised button-like structure called the hypopygium, which when unfolded reveals a pair of genital claspers. In the female fly there is no such button-like protuberance.

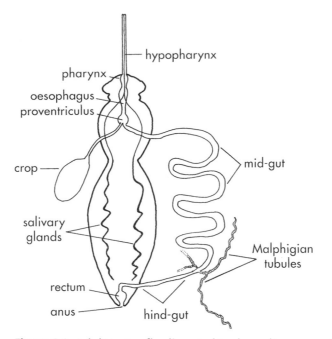

Figure 8.3 Adult tsetse-fly, dissected to show diagrammatically the alimentary canal and paired thread-like salivary glands.

8.1.1   Alimentary canal of the adult fly

Knowing the morphology of the alimentary canal and associated salivary glands (Fig. 8.3) is essential for understanding the life cycle of trypanosomes in the tsetse-fly.

The food channel in the proboscis leads to the pharynx and then the oesophagus, which has a slender duct leading to the oesophageal diverticulum, commonly called the crop. Just behind the oesophagus is a bulbous structure termed the *proventriculus*. The distal end of the proventriculus marks the end of the fore-gut and beginning of the mid-gut, which in tsetse-flies is very long and convoluted. The *peritrophic membrane* is secreted by epithelial cells in the anterior part of the proventriculus and has an important role in the cyclical development of sleeping sickness trypanosomes in the tsetse-fly. When first produced the peritrophic membrane is very delicate, soft and almost fluid, but as it passes through the gut it hardens to form a thin but relatively tough sleeve which lines the entire length of the mid-gut.

The junction of the four Malpighian tubules separates the mid-gut from the hind-gut.

The thread-like paired salivary glands originating in the head of a tsetse-fly are enormously long, very convoluted and stretch back almost to the end of the abdomen. Anteriorly, ducts from both glands unite in the head to form the common salivary duct; this passes through the hypopharynx, which runs down the centre of the proboscis.

8.2   Life cycle

8.2.1   Feeding and reproduction

Both male and female tsetse-flies bite people, a large variety of domesticated and wild mammals, and sometimes reptiles and birds. No species of tsetse feeds exclusively on one type of host but most show definite host preferences, often associated with host availability. For example, in East Africa *Glossina swynnertoni* feeds mainly on wild pigs and *G. morsitans* on wild and domesticated bovids as well as on wild pigs, whereas in West Africa *G. morsitans* feeds mainly on warthogs. In East Africa *G. pallidipes* feeds principally on wild bovids, while in West Africa *G. palpalis* feeds predominantly on reptiles and humans, and in West Africa *G. tachinoides* feeds on humans and bovids, but in southern Nigeria it feeds predominantly on domestic pigs. Tsetse-flies blood-feed about every 2–3 days, although in cool humid conditions it may be about every 10 days. Feeding is restricted to the daytime and vision, as well as olfactory cues emanating from host breath and urine, are important in host location, dark moving objects being particularly attractive. On pale-skinned people, such as Caucasians, tsetse-flies often bite through dark clothing such as socks, trousers and shorts in preference to settling on the skin. During feeding blood sucked up the proboscis passes to the crop and later to the mid-gut, where digestion proceeds.

The different types of flies so far described in this book lay eggs; in marked contrast tsetse-flies do not, but instead they deposit larvae, one at a time (i.e. they are *larviparous*). Adults of *Sarcophaga* and *Wohlfahrtia* also deposit larvae, not eggs (Chapter 10).

After females have mated and taken a blood-meal a single egg in one of the two ovaries completes maturation. It then passes down the common oviduct into the uterus, where it is fertilized by sperm from the paired spermathecae. The egg hatches within the uterus after about 3–4 days. The uterus is supplied with a conspicuous pair of branched secretory accessory glands called the *milk glands* (Fig. 8.4). Secretions from these glands provide the larva with all the food it needs for growth and development. The larva passes through three instars in the female. Regular blood-meals must be taken for a continuous and adequate provision of nutrient fluid from the milk glands. If the fly is unable to feed, the larva may fail to complete its development and as a consequence be 'aborted'.

Larval development takes about 9 days, by which time the third and final instar larva is 5–7 mm long. It is creamy white and composed of 12 visible segments, the last of which bears a pair of prominent dark protuberances called the *polypneustic lobes* (Fig. 8.5b), which are respiratory structures.

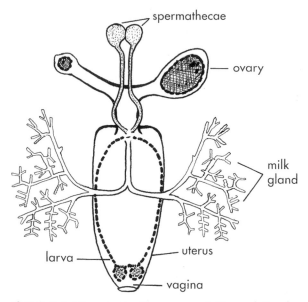

Figure 8.4 Diagrammatic representation of the female reproductive system of a tsetse-fly with a full-grown larva in the uterus.

A female containing a fully developed larva is easily recognized because the fly's abdomen is enlarged and stretched, i.e. the fly is obviously 'pregnant'. Furthermore, the black larval polypneustic lobes can be seen through her abdominal integument.

The mature third-instar larva wriggles out, posterior end first, from the genital orifice; thus birth can be termed a 'breech case'. Females select shaded sites for larviposition. The larva is deposited on loose friable soil, sand or humus, frequently underneath bushes, trees, logs, rocks, between buttress-roots of trees, in sandy riverbeds, in animal burrows and even in rot-holes in trees which may be 4–5 m above the ground. Immediately after the larva is deposited it commences to bury itself under 2–5 cm of soil. After about 15 minutes the third-instar larval skin contracts and hardens to form a reddish brown or dark brown barrel-shaped *puparium* which measures about 5–8 mm, and has, like the larva, distinct polypneustic lobes (Fig. 8.5a). Within this puparial case the larva pupates.

Duration of the puparial period is comparatively long, usually 4–5 weeks, but at high temperatures (30 °C) it may be completed within 3 weeks, and conversely at low temperatures (20 °C) prolonged to 7 weeks. After completion of puparial development the fly emerges from the puparium, forces its way to the surface of the ground and flies away.

During larval development in the female the tsetse-fly feeds every 2–3 days. The first larva is deposited about 16–20 days after the female has emerged from the puparium; thereafter, if food is plentiful, a larva is

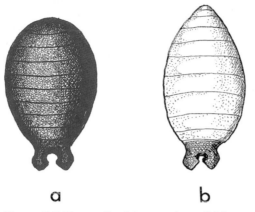

a   b

Figure 8.5 Tsetse-fly: (a) puparium; (b) larva.

deposited about every 9–12 days. In the laboratory female tsetse-flies have produced up to 20 offspring, but the average is nearer 5–8. Breeding generally continues throughout the year but in very humid conditions reproduction may be diminished. Maximum population size is usually reached at the end of the rainy season. The population diminishes in the dry season, when suitable areas of refuge for adult flies and suitable larviposition sites may become restricted and localized.

8.2.2   Adult behaviour
Knowledge of the behaviour of tsetse-flies is essential for understanding their role in the transmission of sleeping sickness, and for the development of effective control strategies.

Blood-engorged tsetse-flies, and unfed hungry flies waiting to feed, spend the night and much of the day resting in dark and usually humid sites. In fact tsetse-flies spend about **23 hours** a day resting on vegetation. During the day favoured resting sites of most species are twigs, branches and trunks of trees and bushes. Flies are not found resting in sites having temperatures above about 36 °C. At night tsetse-flies rest mainly on the upper surfaces of leaves. Accurate knowledge of resting sites may be required for control measures. For example, the height at which adults rest on trees determines the height at which they need to be sprayed with insecticides. Most species in fact rest below 4 m; in Nigeria 50% of *G. palpalis* and *G. morsitans* commonly rest between ground level and 30 cm.

Based on their morphology, ecology, karyotype and behaviour, tsetse-flies can be separated into the following three main groups

Fusca group (forest flies)
The *fusca* group contains 13 species of *Glossina*, all of which are large (10.5–15.5 mm). They are forest flies (except *G. longipennis*, which occurs in arid

areas of East Africa) and most are restricted to the equatorial forests of West and western Central Africa. For example, *G. fusca* occurs mainly in relict forests of West and Central Africa, whereas *G. brevipalpis* is found in secondary forests of East Africa.

The *fusca* group rarely feeds on people and none of the species is a vector of sleeping sickness.

Morsitans group (savannah flies)

Five species are included within the *morsitans* group. They are medium-sized insects, 7.5–11 mm long, and typically inhabit savannah regions of Africa, which may extend from the coast, or the edges of forests, to dry semidesert regions. *Glossina morsitans* occupies the savannah regions of West, Central and East Africa, whereas *G. pallidipes* is found in savannahs of East Africa and parts of southern Africa, and *G. swynnertoni* is restricted to the savannahs of a very limited area of East Africa. Within savannahs *G. morsitans* and *G. pallidipes* occur in areas ranging from wooded sites at the edges of forests to dry thicket vegetation of arid zones, whereas *G. swynnertoni* is restricted mainly to relatively dry thicket country.

All three species mentioned above are vectors of sleeping sickness. The most important is *G. morsitans*, although this species is not a vector in West Africa.

Palpalis group (riverine and forest flies)

Nine tsetse species and subspecies are included in the *palpalis* group, the smallest being about 6.5 mm in length and the largest 11 mm. They are essentially flies inhabiting wetter types of vegetation, such as forests, luxuriant scrub and vegetation growing along rivers and shores of lakes. *Glossina palpalis* inhabits riverine vegetation bordering rivers and lakes, mangrove swamps and forested areas, and occurs throughout most of West Africa, down the western part of the continent to Angola. *Glossina fuscipes*, which is closely related to *G. palpalis*, occurs mainly in Central Africa but extends its range to the western areas of East Africa. *Glossina tachinoides* is a riverine species found near streams and rivers in wet humid coastal areas, through wooded savannah regions to the riverine vegetation of very dry savannah areas. It is found mainly in West and Central Africa but also occurs in parts of Ethiopia and Sudan. All these species are vectors of sleeping sickness.

8.3   Medical importance

All *Glossina* species are potential vectors of African trypanosomiasis to humans. However, relatively few species of tsetse-flies are natural vectors because many species rarely, if ever, feed on people. It is the behaviour of adult tsetse-flies and the degree of fly–human contact, and in the case of

Rhodesian sleeping sickness also the degree of vector contact with reservoir hosts of the trypanosomes, that establishes whether a tsetse-fly is a vector.

Sleeping sickness has a patchy distribution over about 10^6 km^2 of land (that is an area greater than the USA) in 36 African countries. About 30 000 new cases are recorded annually, but the WHO estimates there are actually about 400 000 new cases a year and 55 000 deaths. There are two sub-species of trypanosomes causing sleeping sickness in humans, namely *Trypanosoma brucei gambiense* and *T. brucei rhodesiense*. These parasites are morphologically indistinguishable but produce different clinical symptoms and have different epidemiologies. The most important vectors of sleeping sickness are *G. palpalis*, *G. fuscipes*, *G. tachinoides*, *G. morsitans* and *G. pallidipes*.

The developmental cycle of *Trypanosoma brucei gambiense* and *T. brucei rhodesiense* in the tsetse-fly is the same, and is as follows. Trypanosomes in the blood are sucked up by male and female tsetse-flies during blood-feeding from an infected person (or, particularly in the case of *T. brucei rhodesiense*, often from a non-human reservoir host). They pass through the oesophagus to the crop, and then, after feeding has ceased, into the *peritrophic tube* lining the mid-gut. About 9–11 days after feeding the trypanosomes penetrate the middle section of the peritrophic membrane and pass across into the space between the membrane and gut, called the *ectoperitrophic space*. Here they multiply, and after 3–9 days the parasites penetrate the anterior softer part of the peritrophic membrane and migrate to the *proventriculus*. From here they pass down the food channel in the proboscis and pass up the hypopharynx or salivary duct to invade the *salivary glands*, where they develop into *epimastigotes* and multiply enormously. About 18–35 days after an infective blood-meal the tsetse-fly becomes infective and metacyclic *trypomastigotes* (metatrypanosomes) are injected into a vertebrate host when the fly feeds. In the 1970s there was evidence for a more direct route in which parasites passed through the peritrophic membrane and gut wall to enter the haemocoel (body cavity) and then invaded the salivary glands. It remains unclear how important this shorter route is.

When determining mature infection rates in tsetse-flies by dissection, any trypanosomes found in the gut or proboscis are ignored as only those in the salivary glands can be *T. brucei gambiense* or *T. brucei rhodesiense* infections, although immature forms occur elsewhere. There are complications, because *T. brucei brucei*, which does not cause sleeping sickness in humans but causes an animal trypanosomiasis commonly called nagana, undergoes a similar cyclical development in the fly. Consequently, the presence of trypomastigotes (metacyclic forms) in the salivary glands does not necessarily indicate the presence of trypanosomes infective to people.

Salivary gland infection rates in tsetse-flies are low, rarely exceeding 0.1%, even in endemic areas.

8.3.1   Gambian sleeping sickness

Gambian sleeping sickness (*T. brucei gambiense*) is a form of the disease that occurs from West Africa through Central Africa to parts of Sudan and southwards to Angola and the Democratic Republic of Congo (formerly Zaire). *Glossina palpalis* and *G. tachinoides* are the most important vectors in West Africa, while *G. fuscipes* is the vector in Central and East Africa. This disease is relatively chronic, with death often not occurring until after many years. Until recently it was considered that there were no natural animal reservoir hosts, but it has been shown that pigs can harbour the parasites; whether they play any role in transmission to humans remains unclear.

Vectors of Gambian sleeping sickness are especially common at watering places, fords across rivers and along lake shores, etc., situations that people frequently visit to collect water or do their washing. As a consequence there may be limited and localized foci of transmission.

8.3.2   Rhodesian sleeping sickness

The causative agent of Rhodesian sleeping sickness, *T. brucei rhodesiense*, causes a more virulent disease than *T. brucei gambiense*, with death occurring after just weeks or months. However, it is not so widespread, being more or less restricted to Tanzania, Malawi, Zambia, Zimbabwe, Mozambique and the northern areas of Lake Victoria in Kenya and Uganda. The most important vectors are *G. morsitans* and *G. pallidipes*, species which feed on a variety of game animals and domestic livestock, especially bovids, in preference to people. These flies often occur in savannah areas thinly populated by humans. Wild animals, especially various bovid species, are important reservoir hosts of *T. brucei rhodesiense*. Around Lake Victoria, *G. fuscipes* is the main vector and cattle are important reservoir hosts. This form of sleeping sickness is a *zoonosis*.

8.4   Control

Because the larva is retained by the female for almost all of its life, and the puparium is buried in the soil, control of tsetse-flies is aimed at the adults.

Many control methods have been directed at tsetse-flies to combat human and animal trypanosomiasis. Formerly in Zimbabwe and some East African countries game animals in selected areas that might provide food for tsetse-flies, or be reservoir hosts of trypanosomiasis, were killed. Such slaughter of animals is no longer acceptable in a world that is increasingly sensitive to the preservation of wildlife.

Distribution and abundance of tsetse-flies is largely determined by types of vegetation. Because of such dependence on vegetation its destruction has in the past achieved considerable success in controlling tsetse-flies, but in many countries this method, at least on a large scale, is no longer ecologically or economically acceptable. However, clearing vegetation for human settlements or felling trees for firewood may also have a similar effect in reducing tsetse-fly populations.

8.4.1   Insecticidal control

There are no problems with insecticide resistance in tsetse-flies. Consequently insecticides such as endosulfan, permethrin, deltamethrin, lambda-cyhalothrin and cypermethrin can be sprayed onto vegetation harbouring adult flies. Success depends on a detailed knowledge of the resting sites of the vectors. For example, it is often possible to restrict spraying of tree trunks up to a height of 1.5 m during the dry season, extending this to 3.5 m in the wet season. Spraying is often done in the dry season so that deposits are not washed from vegetation by rainfall and applications can remain effective for at least 2–3 months.

Alternatively ultra-low-volume (ULV) aerial insecticidal spraying using fixed-wing aircraft can be undertaken. But because the droplets are too small to produce a persistent deposit on vegetation, and because the tsetse population is in the soil as puparia for 4–5 weeks, repeated sprayings are required, for example five or six times at 12- to 18-day intervals.

Although indiscriminate and repeated spraying of residual insecticides can have disastrous effects on the local fauna it should be appreciated that application rates of insecticides for tsetse-fly control are often very low (e.g. 20 mg active ingredient of deltamethrin per hectare). Furthermore, after flies have been controlled and spraying stopped, many, if not all, of the wildlife may revert to their original population numbers. Intensive bush clearance for tsetse-fly control can also greatly diminish the numbers and variety of the local fauna, but in this instance the change may be permanent, especially if people occupy and farm areas that were originally scrub or forest.

Near Lake Victoria, Uganda, where sleeping sickness is caused by *T. b. rhodesiense*, spraying or dipping cattle with pyrethroid insecticides has helped reduce the number of human cases because the vector, *G. fuscipes*, feeds on cattle as well as humans and thus is killed while feeding on this host.

8.4.2   Targets and traps

Although insecticidal spraying is still used to control tsetse-flies there is increasing use of more environmentally friendly, and often cheaper, techniques such as employing targets and traps to control them. Targets, sometimes called screens, made of dark blue or black cloth about 1 m² and impregnated with pyrethroids (e.g. deltamethrin, lambda-cyhalothrin or cypermethrin) are mounted on poles or hung from trees. They are very attractive to tsetse-flies and kill adult flies settling on their surfaces. Reimpregnation is needed every 3–4 months.

Biconical or pyramidal traps made partially from dark blue or black material can be remarkably effective in trapping flies. Because some flies fail to enter traps they are often impregnated with pyrethroid insecticides to kill those resting on the outside. Traps should be retreated with insecticides about every 6–10 months. Location of both screens and traps is critical for efficient control; generally open and sunny sites are best as they offer good visibility for the flies.

It has been estimated that a 5 km^2 village would need about 50 traps or screens to effectively reduce tsetse-flies in the immediate area. The employment of so many traps can be expensive because they require initial financial outlay and maintenance.

8.4.3   Genetic control

Although genetic control, mainly using sterile male release methods (see p. 25), has sometimes been successful against tsetse-fly vectors of cattle trypanosomiasis, even leading to the eradication of flies from Zanzibar, the approach is not generally feasible for sustained control of human sleeping sickness. An exception might be to use genetic methods to eradicate residual populations of flies left locally after cheaper control methods (e.g. insecticides, traps) have drastically reduced their numbers but failed to achieve eradication.

Further reading

Buxton, P. A. (1955) *The Natural History of Tsetse-Flies: an Account of the Biology of the Genus* Glossina (*Diptera*). London School of Hygiene and Tropical Medicine Memoir, 10. London: H. K. Lewis

Colvin, J. and Gibson, G. (1992) Host-seeking behavior and management of tsetse. *Annual Review of Entomology*, **37**, 21–40.

Fèvre, E. M., Picozzi, K., Jannin, J., Welburn, S. C. and Maudlin, I. (2006) Human African trypanosomiasis: epidemiology and control. *Advances in Parasitology*, **61**, 168–221.

Ford, J. (1971) *The Role of Trypanosomiases in African Ecology: a Study of the Tsetse Fly Problem*. Oxford: Clarendon Press.

Gooding, R. H. and Krafsur, E. S. (2005) Tsetse genetics: contributions to the biology, systematics, and control of tsetse flies. *Annual Review of Entomology*, **50**, 101–23.

Green, C. H. (1994) Bait methods for tsetse fly control. *Advances in Parasitology*, **34**, 229–91.

Jannin, J. and Cattand, P. (2004) Treatment and control of human African trypanosomiasis. *Current Opinions in Infectious Diseases*, **17**, 565–71.

Jordan, A. M. (1986) *Trypanosomiasis Control and African Rural Development*. London: Longman.

Jordan, A. M. (1989) Man and changing patterns of the African trypanosomiases. In *Demography and Vector-Borne Diseases*, ed. M. W. Service. Boca Raton, FL: CRC Press, pp. 47–58.

Jordan, A. M. (1993) Tsetse-flies (Glossinidae). In *Medical Insects and Arachnids*, ed. R. P. Lane and R. W. Crosskey. London: Chapman and Hall, pp. 333–88.

Leak, S. G. A. (1999) *Tsetse Biology and Ecology: Their Role in the Epidemiology and Control of Trypanosomiasis*. Wallingford: CABI.

Lutumba, P. (2005) Trypanosomiasis control, Democratic Republic of Congo, 1993–2003. *Emerging Infectious Diseases*, **11**, 1382–9.

Maudlin, I. (2006) African trypanosomiasis. *Annals of Tropical Medicine and Parasitology*, **100**, 679–701.

Maudlin, I., Holmes, P. H. and Miles. M. A. (eds.) (2004) *The Trypanosomiases*. Wallingford: CABI.

Molyneux, D. H. (2001) Sterile insect release and trypanosomiasis control: a plea for realism. *Trends in Parasitology*, **17**, 413–4.

Mulligan, H. W. (ed.) (1970) *The African Trypanosomiases*. London: Allen and Unwin.

Nash, T. A. M. (1969) *Africa's Bane: the Tsetse Fly*. London: Collins.

Torr, S. J., Hargrove, J. W. and Vale, G. A. (2005) Towards a rational policy for dealing with tsetse. *Trends in Parasitology*, **21**, 537–41.

World Health Organization (1998) Control and surveillance of African trypanosomiasis. *World Health Organization Technical Report Series*, **881**, 1–113.

World Heath Organization (2003) *Report of the Scientific Working Group Meeting on African Trypanosomiasis (Sleeping Sickness)*. TDR/SWG/01. Geneva: WHO.

9

House-flies and stable-flies (Muscidae) and latrine-flies (Fanniidae)

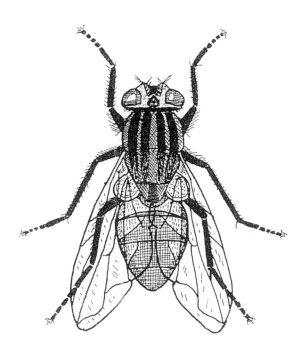

There are 18 000 or so species of true flies (sometimes called calyptrate Diptera), and this vast number includes several medically important species in the families Glossinidae (Chapter 8), Muscidae and Fanniidae, the last two of which are described in this chapter.

The Muscidae contains about 4200 species of flies in 170 genera. The medically most important are the common house-fly (*Musca domestica*), the greater house-fly (*Muscina stabulans*) and the stable-fly (*Stomoxys calcitrans*), all having a more or less worldwide distribution. House-flies can be vectors of helminths, faecal bacteria, protozoans and viruses, resulting in the spread of enteric diseases (e.g. dysenteries and typhoids). The stable-fly can cause a biting nuisance.

The family Fanniidae comprises about 280 species in four genera but only species in the genus *Fannia*, such as *F. canicularis* (lesser house-fly) and *F. scalaris* (latrine fly), are of medical importance, and like house-flies they can transmit various pathogens to humans.

9.1   The common house-fly (*Musca domestica*)

9.1.1   External morphology

There are about 70 species of flies in the genus *Musca*. The most common is *M. domestica*, the house-fly, which is almost worldwide but is least common in Africa, where it is largely replaced by two subspecies (*M. domestica curviforceps* and *M. domestica calleva*). Other important species are (1) the bazaar-fly (*Musca sorbens*), which can be a great nuisance in Africa, Asia and the Pacific; (2) the notorious troublesome bush-fly of Australia, namely *M. vetustissima*; and (3) the face-fly (*M. autumnalis*), which is a pest in both the Old and New Worlds. The appearance and biology of these *Musca* species are very similar. The morphology and biology of the house-fly (*M. domestica*) are described here.

House-flies are medium-sized ***non-metallic*** flies about 6–9 mm long, varying in colour from light to dark grey with some darker markings. They have four broadish black longitudinal stripes on the dorsal surface of the thorax (Fig. 9.1a, Plate 10). The antennae, which are not easily seen, are concealed in depressions on the front of the face. Each antenna consists of three segments, the distal and largest of which is cylindrical and has a prominent hair, called an ***arista***, which has hairs on ***both*** sides.

The mouthparts (proboscis) are specially adapted for sucking up fluid or semifluid foods. When not in use they are partially withdrawn into the head capsule (Fig. 9.2a), but are extended vertically downwards in a telescopic fashion when the fly feeds (Fig. 9.2b). The proboscis ends in a pair of oval-shaped fleshy ***labella***, having very fine channels called ***pseudotracheae*** through which fluids are sucked up. House-flies feed on many types of substances, including almost all food of humans, rotting vegetables, carcasses, excreta and vomit – in fact almost any organic material.

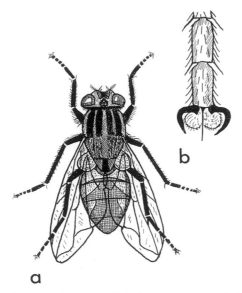

Figure 9.1 House-fly (*Musca domestica*): (a) adult fly; (b) terminal tarsal segments, showing paired claws, paired large pulvilli and a single bristle-like empodium.

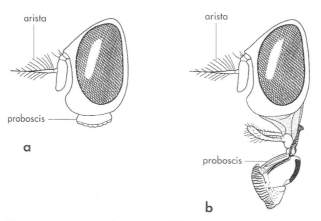

Figure 9.2 *Musca domestica*: (a) lateral view of head of adult house-fly with proboscis retracted; (b) proboscis extended for feeding.

The method of feeding differs according to the physical state of the food. For example, for thin fluids, such as milk and beer, the labella are placed in contact with food, which is then sucked up through small openings in the pseudotracheae. When feeding on semisolids such as excreta, sputum and nasal discharges, the labella are completely everted and food is sucked up directly into the food channel. When flies feed on more solid materials such as sugar lumps, dried blood, cheese and cooked meats, the labella are everted and minute prestomal teeth surrounding the food channel are exposed and scrape the solid food. The fly then moistens small food particles with either saliva or the regurgitated contents of its crop so that food can be sucked up. This latter type of feeding is clearly conducive to the spread of a variety of pathogens.

The wings of the house-fly have *vein 4* bending up sharply to join the costa (the thick vein along the front edge of the wing) close to vein 3 (Fig. 9.3). This is an important identification character which helps distinguish *Musca* species from other rather similar flies. All three pairs of legs end in paired claws and a pair of fleshy pad-like structures called the *pulvilli*, which are supplied with glandular hairs (Fig. 9.1b). These sticky hairs enable the fly to adhere to very smooth surfaces, such as windows. They are also responsible for the fly picking up pathogens when it visits excreta, septic wounds, rubbish dumps, etc. There are four visible greyish abdominal segments, which are usually partially obscured by the wings.

9.1.2   Life cycle

Female *Musca domestica* lay their eggs on decomposing materials such as animal manure, poultry dung, urine-contaminated bedding, carcasses, decomposing organic materials found in rubbish dumps, household garbage and waste foods from kitchens. Some 75–150 eggs are deposited together, or in separate batches. A fly may lay eggs 5–10 times in her lifetime, sometimes totalling up to 1000 eggs. Eggs are creamy-white, 1–1.2 mm long, and distinctly concave dorsally, giving them a banana-shaped appearance (Fig. 9.4a). They can hatch after only 10–16 hours, but this period is extended in cool weather. Hatching is accomplished by the strip of eggshell between parallel ridges on the dorsal concave surface lifting up, and partially detaching itself from the rest of the egg. Eggs cannot withstand desiccation and die if they dry out. Neither can they tolerate extremes of temperatures, most dying after exposure to temperatures below 15 °C or above 40 °C.

Larvae, known as *maggots*, have a small head followed by an 11-segmented cylindrical body (Fig. 9.4b). At the pointed head end a pair of blackish small curved mouthhooks can be seen beneath the integument and are continued as the cephalopharyngeal skeleton in the anterior thoracic segments. At the posterior end of the body there is a pair of conspicuous *spiracles* shaped like a *letter D*. Each spiracle has a complete and thick outer wall called the *peritreme*, which encloses three very sinuous spiracular slits (Fig. 9.4c).

Larvae feed on liquids from decomposing organic material. There are three larval instars. Mature larvae measure about 8–14 mm, the final size depending on environmental conditions, especially the amount of

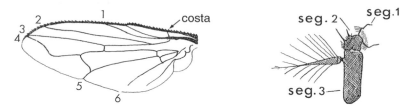
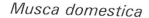
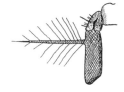
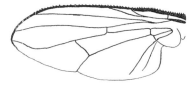

Musca domestica

Muscina stabulans

Fannia canicularis

Stomoxys calcitrans

Figure 9.3 Wing venation and antennae characteristic of the genera *Musca, Muscina, Fannia* and *Stomoxys*. Note endings of veins 3 and 4.

available food. The speed of larval development depends on the abundance of food and temperature. It may be completed in just 3–5 days, but under less favourable conditions 7–10 days are needed, and in cool weather development may extend to about 24 days.

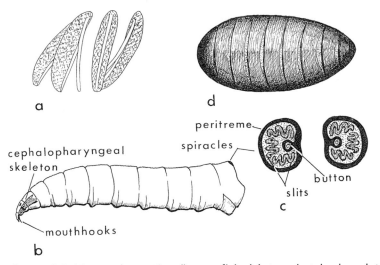

Figure 9.4 *Musca domestica* (house-fly): (a) two hatched and two unhatched eggs; (b) final-instar larva; (c) posterior larval spiracles showing their D-shape; (d) puparium.

Prior to pupation third-instar larvae often move to drier ground. Sometimes, however, the periphery of breeding places, such as rubbish dumps, may be sufficiently dry for larvae to pupate there; pupation may also occur in the dry soil underneath larval habitats. Pupation begins with the larval skin contracting, hardening and turning dark brown, resulting in a barrel-shaped structure about 6 mm long, called a *puparium*. Close examination shows this is segmented (Fig. 9.4d). The puparium is commonly called the pupa but technically this is incorrect because the actual pupa is formed within the protective shell of the puparial case. A puparium is also formed by tsetse-flies (Chapter 8) and by myiasis-causing flies (Chapter 10). The puparial stage lasts about 3–5 days in warm weather but 7–14 days during cooler periods.

Developmental time from egg to adult (Fig. 9.5) is about 49 days at 16 °C, 25 days at 20 °C, 16 days at 25 °C, 10–12 days at 30 °C and 6–7 days at 33 °C. Very occasionally the period can be less than 7 days. Immature development ceases at temperatures below 12 °C, and 45 °C is lethal to the eggs, larvae and puparia. In temperate areas a varying, but usually small, proportion of house-flies survive through the winter as puparia, but more frequently they overwinter as hibernating adults.

The adult fly escapes from its puparial case by pushing off its anterior end and crawling out, and after a short period it flies away.

Adult *Musca domestica* generally avoid direct sunlight, preferring to shelter in buildings inhabited by people or their animals. House-flies and related flies, as well as the calliphorid flies (Chapter 10), are often called

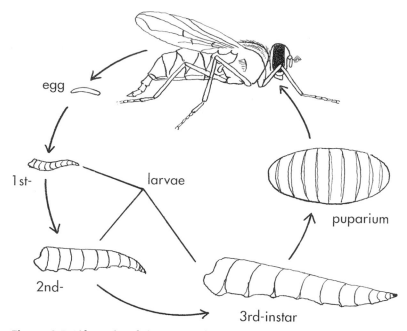

egg

1st-   larvae

puparium

2nd-

3rd-instar

Figure 9.5 Life cycle of the house-fly (*Musca domestica*), which is also typical of other muscid and calliphorid flies (see Chapter 10).

domestic or **synanthropic** flies, because of their close association with humans and their homes. House-flies defecate at random, and frequently regurgitate their food, resulting in unsightly 'fly spots'. Adults tend to stay within 500 m of their breeding sites, but may fly 1–5 km or sometimes even further.

Musca sorbens is widely distributed in Africa and parts of Asia, where it is a greater pest than *M. domestica* because adults more commonly settle around the eyes, on sweaty skin, suppurating wounds, and other body secretions than does the common house-fly.

9.1.3   Medical importance
House-flies can transmit a large number of infections to humans because of their habits of visiting, almost indiscriminately, faeces and other unhygienic matter and people's food. Pathogens can be transmitted by three possible routes:

(1)   By flies' contaminated feet, body hairs and mouthparts. Most pathogens, though, remain viable on the fly for less than 24 hours, and there are usually insufficient numbers to cause a direct infection, except possibly with *Shigella*. However, if pathogens are first transferred to food they may then multiply sufficiently to reach the level of an infective dose.

(2) By flies vomiting on food during feeding, which they do frequently.
(3) By defecation, which often occurs on food. This is probably the most important method of transmission.

Over 100 different pathogens have been recorded from house-flies, and it has been reported that 65 of them are transmitted to people. With the exception of *Thelazia* species (eye-worms) transmission is mechanical, that is the fly acts just as a physical carrier.

House-flies can transmit *viruses* of polio, Coxsackie and infectious hepatitis; *rickettsiae* of Q fever (*Coxiella burnetii*); *bacteria* such as anthrax, *Campylobacter*, cholera (*Vibrio cholerae*), *Shigella* and *Salmonella*, *Escherichia coli*, *Staphylococcus aureus*; *spirochaetes* of yaws (*Treponema pertenue*); *protozoans* including *Entamoeba*, *Cryptosporidium* and *Giardia*. In addition, house-flies can carry eggs of a variety of helminths, for example *Taenia*, *Ancyostoma*, *Dipylidium*, *Diphyllobothrium*, *Enterobius*, *Trichuris* and *Ascaris*. Eye-worms (*Thelazia* species) are rather rare infections of the eye and are transmitted biologically by *Musca* species.

It seems that because house-flies do not commonly rest on or near the eyes they are unlikely to be important vectors of trachoma (*Chlamydia trachomatis*). In marked contrast adults of *Musca sorbens*, the bazaar-fly, frequently settle on or around the eyes and have been shown to play an important role in the transmission of trachoma. In trachoma-endemic areas therefore their breeding sites, mainly human faeces, should be greatly reduced.

In the tropical Americas house-flies can also carry the eggs of *Dermatobia hominis*, a myiasis-producing fly (Chapter 10).

Larvae of house-flies have occasionally been recorded causing urogenital and traumatic myiasis, and more rarely aural and nasopharyngeal myiasis. If food infected with fly maggots is eaten, then they may be passed more or less intact in the excreta, often causing considerable alarm and surprise. There is, however, no true intestinal myiasis in humans (see Chapter 10 for accounts of myiasis).

Although house-flies are potential vectors of many pathogens to humans it may be difficult to assess their importance in disease transmission. Their role in the spread of disease is often circumstantial: for example, seasonal increase of fly abundance associated with outbreaks of diarrhoeal diseases.

A classic demonstration of the association between flies and disease was in two groups of Texan towns in 1946 and 1947. One group was sprayed with DDT to destroy house-flies, and this was accompanied by a reduction in acute diarrhoeal diseases and deaths in children due to *Shigella*, although *Salmonella* infections remained the same. The unsprayed town did not have any reduction in *Shigella* infections. Later the unsprayed towns were sprayed and the towns previously sprayed left unsprayed.

The incidence of dysentery in the two groups was now reversed. In 1988 control of house-flies by attractant traps in an Israeli army camp resulted in a reduction in *Shigella* infections and also apparently enterotoxigenic *Escherichia coli* (ETEC) infections. More recently, in 1995 and 1996, when breeding sites in Pakistani villages were sprayed with insecticides fly populations decreased by about 97% and the incidence of childhood diarrhoea decreased by 23%. Similarly in The Gambia, ultra-low-volume (ULV) spraying with deltamethrin in 1997 and 1998 reduced fly populations by about 75% and there were 75% fewer new cases of trachoma; diarrhoea in children was reduced by 25%. There is now considerable evidence that house-flies contribute to ill health.

In 2004 investigations in a Misurata city hospital in Libya demonstrated that house-flies were carrying antibiotic-resistant bacteria, including methicillin-resistant *Staphylococcus aureus* (MRSA).

9.1.4   Control
Control methods can be divided conveniently into three categories:

(1)   physical and mechanical control
(2)   environmental sanitation
(3)   insecticidal control

Physical and mechanical control
Flies can sometimes be prevented from entering buildings by screening windows and other openings. Plastic material is preferable because metal screening is liable to corrode. Mesh size of 3–4 strands per centimetre (i.e. 2–3 openings per cm) will exclude house-flies from buildings without unduly decreasing air circulation or light. Screening can be costly, but may be worthwhile in hospitals and restaurants. Screening should be periodically inspected and tears mended. Although screening can reduce fly nuisance, flies will continue to breed locally and enter unscreened houses.

Air currents, such as the air barriers found in the entrances of some shops, and fans mounted over doorways, may reduce the number of flies entering premises. Placing curtains of vertical, often coloured, strips of plastic or beading in doorways also helps to keep out flies. Restaurants, food stores and hospitals often mount ultraviolet light-traps on walls to attract flies, which are then killed by an electric grid. Commercially available sticky tapes ('fly-papers'), incorporating sugar as an attractant, can be relatively effective, although unsightly, in catching flies.

Environmental sanitation
Environmental sanitation aims at reducing house-fly populations by minimizing their breeding places. For example, domestic refuse and garbage should be placed either in strong plastic bags with the openings tightly

closed, or in dustbins with tight-fitting lids. When possible there should also be regular refuse collections, preferably twice a week in warm countries, to prevent eggs laid in the garbage developing into adults. If household refuse cannot be collected it should be burnt or buried. Unhygienic rubbish dumps, so commonly found in towns and villages, provide ideal breeding places for house-flies, and should be removed. Instead refuse should be placed in pits and covered, daily if possible, with a 15-cm layer of earth; when the pits are more or less full they must be covered with 60 cm of compacted earth. This depth of final fill is required to prevent rodents burrowing into the rubbish pits.

Insecticidal control
Insecticide resistance develops very quickly in house-flies, and resistance to various insecticides occurs worldwide. In the absence of resistance insecticides such as diazinon, fenitrothion, malathion, pirimiphos-methyl, deltamethrin, permethrin, cypermethrin and etofenprox can be used.

Larvicides Insecticides can be directed against the larvae by spraying the insides of dustbins and refuse and garbage heaps, manure piles and other breeding sites. Insect growth regulators (IGRs) such as diflubenzuron, cryomazine or pyriproxyfen are also used. Usually large volumes (0.5–5 litres/m^2) are needed to penetrate the upper 10–15 cm of breeding sites to reach the larvae.

Spraying against adults Commercial aerosol spray cans or hand sprayers can be used indoors to spray knock-down insecticides to give an immediate kill of adult flies. Suitable insecticides include 2–4% malathion, 1% naled, 2% pirimiphos-methyl, 0.1–0.5% fenchlorphos, or 0.1–0.2% permethrin, cypermethrin or deltamethrin. Care must be taken not to contaminate food with insecticides. Aerosol applications and space-spraying have virtually no residual effects: consequently treatments have to be repeated, and this can be costly. Furthermore this approach does little to alleviate the source of the fly nuisance.

Outdoor aerosol applications or aerial ULV spraying with pyrethroids can give effective control in and around dairies, farms, markets and recreational areas. Usually, however, only temporary relief from flies is achieved because outdoor spraying does not usually kill indoor-resting flies, prevent their reinvasion from outside the treated area, or prevent the emergence of new flies from breeding places.

Residual spraying Flies may also be controlled by spraying indoor walls, ceilings, doors, etc. with 1–2 g/m^2 of malathion, fenitrothion or pirimiphos-methyl; or 0.4–1.0 g/m^2 diazinon or naled; or 0.1–0.4 g/m^2 bendiocarb or propoxur; or 0.025–0.1 g/m^2 cypermethrin or permethrin; or 0.01–0.15 g/m^2 deltamethrin. These residual insecticides should remain effective for 1–2 months, but this depends on local conditions. Outside walls of houses and cattle sheds can also be sprayed with residual insecticides, but their duration of effectiveness will depend on whether the surface deposit is rubbed off, or washed off by rain.

The outside of dustbins and adjacent walls should also be sprayed to deter gravid (egg-laden) flies from ovipositing in dustbins and other nearby breeding places.

Insecticidal cords   Cords or rope strips soaked in insecticides such as diazinon, dimethoate, malathion, propoxur, cypermethrin or permethrin and dyed, preferably red, to alert people that they are impregnated with insecticides, can be hung up in dairies and other premises. They remain effective in killing flies resting on them for 2–6 months, depending on the insecticide and dosage.

Toxic baits   Dry baits consist of sugar mixed with bran or crushed corncobs, or some other inert carrier treated with insecticides. They provide an attractive but lethal solid bait which can be scattered on floors or placed in trays; they need to be replaced about every two days. Liquid baits commonly comprise 10% sugar solution and an insecticide. This is placed in a glass bottle inverted over a saucer-like receptacle so that as the bait evaporates more flows in from a reservoir, as in automatic feeders for poultry. The most commonly used insecticides in both dry and liquid baits are propoxur, diazinon, naled, dimethoate and trichlorphon. Newer insecticidal compounds are spinosad (actinomycete) and imidacloprid (nicotinoid). Commercially prepared dry or liquid baits are sometimes available. The house-fly sex pheromone muscalure can be added to baits to make them more attractive.

Viscous paint-on baits comprising an insecticide, sugar and a binder can be painted on a variety of vertical or horizontal surfaces, and last for 1–2 months or longer. Baits can sometimes quickly reduce fly populations.

9.2   The greater house-fly (*Muscina stabulans*)

9.2.1   External morphology

Muscina stabulans has a worldwide distribution and is commonly called the greater house-fly, or false stable-fly. Adults are about 7–10 mm long, slightly larger than house-flies. They are distinguished from both *Musca* and *Fannia* species because *vein 4* of the wing curves slightly, but distinctly, upwards towards vein 3 (Fig. 9.3), and from *Fannia*, but not *Musca*, by having hairs on **both** the upper and lower sides of the arista (Fig. 9.3).

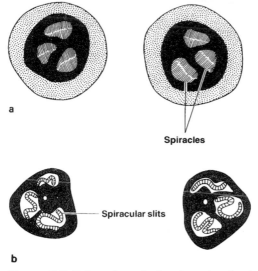

Spiracles

Spiracular slits

Figure 9.6 Pairs of posterior larval spiracles: (a) *Muscina stabulans* (greater house-fly); (b) *Stomoxys calcitrans* (stable-fly).

9.2.2   Life cycle

Females scatter their eggs mainly on decaying organic matter and human and animal excreta. Eggs hatch after 1–2 days and the resultant larvae resemble the maggot-shaped larvae of the house-fly, but can be distinguished by having almost *circular* posterior *spiracles*, not D-shaped as in *M. domestica*. Furthermore, the *peritreme*, which is very wide, encircles three *crescent-shaped* spiracular slits (Fig. 9.6a). Young larvae are omnivorous scavengers but towards the end of the larval period they become predaceous, feeding on other fly larvae. The brown puparium is similar to that of *M. domestica* and its duration is about 1–2 weeks. The life cycle lasts about 4–6 weeks, although in warm weather it may be reduced to 20–25 days.

Adults feed and behave much like adults of *M. domestica*, and may enter buildings.

9.2.3   Medical importance

Adults frequently defecate on food, but the exact role of *M. stabulans* as a disease vector remains undetermined. However, many of the pathogens transmitted by the house-fly are probably also spread by this species. Larvae, like those of *Fannia* and *Musca*, are occasionally found in stools, but there is no true intestinal myiasis.

9.2.4   Control

Control measures are very similar to those for the house-fly.

Figure 9.7 Lateral view of the head of *Stomoxys calcitrans* (stable-fly) showing the forward-projecting proboscis.

9.3   The stable-fly (*Stomoxys calcitrans*)

9.3.1   External morphology

Stable-flies have a worldwide distribution. They are sometimes known as biting house-flies or, in the USA, as dog-flies (because they commonly bite dogs). The most common species is *Stomoxys calcitrans*. Adults have four black longitudinal stripes on a dark grey thorax and are about the size (5–6 mm) of house-flies, which they superficially resemble. They are, however, easily separated from *Musca, Fannia* and *Muscina* by their conspicuous forward-projecting, ***rigid proboscis*** (Fig. 9.7). Wing venation resembles that of *Muscina*, but the **arista** of the third antennal segment differs from both *Musca* and *Muscina* in having hairs **only** on the upper side (Fig. 9.3).

In Africa adults could be confused with tsetse-flies, which also have a forward-projecting proboscis, but *Stomoxys calcitrans* is a smaller fly. Also, when at rest its wings are not placed completely over the body in a closed scissor-like fashion, as in tsetse-flies, but are kept apart as in house-flies. Furthermore, unlike tsetse-flies, there is no enclosed hatchet cell in the wings.

9.3.2   Life cycle

Both males and females take blood-meals from wild and domesticated animals, including cattle, horses, pigs and dogs; they also feed on humans, especially if their preferred hosts are absent or scarce. During feeding the forward-projecting proboscis is swung downwards and the skin penetrated; bites can be painful, and most are on the legs. In hot weather blood-meals are digested within 12–24 hours and adults feed about every 1–3 days, but in cooler conditions blood digestion takes 2–4 days or longer, and feeding occurs every 5–10 days. ***Biting*** is restricted to the daytime and occurs both in bright sunshine and in cloudy weather. Although most biting is outdoors stable-flies will also enter houses to feed. Adults are mostly encountered in and around farms or where horses are kept, and consequently they are more common in rural areas than in towns.

Their creamy-white eggs resemble those of house-flies and are usually laid in batches of less than 20, but sometimes as many as 50–200 may be laid together, so that 100–800 eggs may be laid in the fly's lifetime. Eggs are usually deposited in horse manure but also in compost pits, decaying and fermenting piles of vegetable matter, weeds, cut grass or hay. Stable-flies very rarely lay their eggs in human or animal faeces, unless these are liberally mixed with hay or straw.

Eggs hatch within 1–5 days and the resultant larvae are creamy-coloured *maggots* which resemble those of the house-fly. However, they can be separated by having the posterior *spiracles* widely separated (Fig. 9.6b), thus differing from those of *Musca* and *Muscina*. Spiracles are approximately round, lack a conspicuous *peritreme*, and the S-shaped spiracular slits are widely separated from each other. Larvae prefer a high degree of moisture for development and are found mostly in wet mixtures of manure and soil or straw, and in vegetable matter in advanced stages of decay. Under optimum conditions the larval period lasts about 6–10 days, but in cooler weather, or when there is a shortage of food, development can be prolonged to 4–5 weeks or more.

Larvae migrate to drier areas and bury themselves in the soil prior to pupation. The puparium is dark brown and resembles that of the house-fly, but can be distinguished from it by having the posterior spiracles widely separated. The puparial stage usually lasts 5–7 days, although in cool weather it may be extended to several weeks. The life cycle from egg-laying to adult emergence takes 12–13 days at 27 °C, but longer at cooler temperatures. Adults may overwinter.

In tropical areas stable-flies breed continuously throughout the year. In more temperate climates they pass the cooler months as larvae or puparia, but sometimes adults survive the winter in warm stables or buildings, feeding intermittently during the cooler months.

9.3.3   Medical importance
Because of their painful bites *both* sexes of stable-flies can be troublesome pests of people, cattle, horses and pets – notably dogs. Adult flies are not regarded as transmitting diseases to humans, and although they can be mechanical vectors of African trypanosomiasis there is little evidence that they play any part in the epidemiology of sleeping sickness. Because stable-flies rarely visit excreta and festering wounds they are much less likely to spread pathogens in the way that house-flies do.

In the tropical Americas eggs of *Dermatobia hominis*, a myiasis-producing fly (Chapter 10), are sometimes attached to adult stable-flies.

9.3.4   Control
Many control methods aimed at house-flies can be applied with some modification to control stable-flies – for example, not allowing piles of manure,

grass cuttings or decaying vegetable matter to accumulate, and burning straw and bedding material. Breeding places can be sprayed with insecticides, but as noted with house-fly control it is often difficult to get insecticides to penetrate deeply to where most larvae are found. Insecticidal spraying of horse stables, animal shelters, barns and other farm buildings can help reduce the numbers of stable-flies.

9.4   The lesser house-fly and the latrine-fly (*Fannia* species)

9.4.1   External morphology

Here the genus *Fannia* is placed in the family Fanniidae, but some place it in the family Muscidae.

Flies of *Fannia* resemble house-flies but are a little smaller (5–7 mm). They are readily distinguished from house-flies by having **vein 4** of the wing almost parallel to vein 3 (Fig. 9.3), whereas in *Musca* it bends upwards and almost touches vein 3 at the wing apex, and in *Muscina* vein 4 also bends upwards but not to such an extent. The **arista** arises from the third antennal segment and is ***completely*** devoid of hairs (Fig. 9.3), whereas in both *Musca* and *Muscina* there are hairs on both upper and lower sides. In most other respects adult flies are similar to house-flies.

Two common species of *Fannia* are of minor medical importance. They are *Fannia canicularis* (the lesser, or little, house-fly), which occurs worldwide and is commonly encountered in houses, and *Fannia scalaris* (the latrine-fly) which has a Holarctic distribution and is uncommon in houses. Adults of *Fannia canicularis* have **three** rather faint dark longitudinal stripes on the thorax, whereas *F. scalaris* has **two** such stripes.

9.4.2   Life cycle

Eggs of the lesser house-fly (*F. canicularis*) resemble those of the house-fly. They are deposited on food, but also on urine-soaked bedding of humans and animals, compost heaps, decaying piles of grass, human and animal excreta and in poultry litter. The latrine-fly (*F. scalaris*) usually lays her eggs in faeces – hence its common name. Larvae of *Fannia* species prefer wetter breeding places than house-flies and are often found in semiliquid material, such as runny faeces.

Eggs hatch after 1–2 days. *Fannia* larvae are very different to the maggot-shaped larvae of *Musca* and are unlikely to be confused with the larvae of other medically important flies. They are flattened dorsoventrally and have many thin but conspicuous fleshy processes arising from the body which bear small spiniform secondary processes. Figure 9.8 shows minor differences between larvae of *F. canicularis* and *F. scalaris*. Larval development takes about 7–12 days but may be prolonged if habitats start to dry out. The puparium is brown and is similar in shape to the larva. After 7–10 days the adult fly emerges. The life cycle often lasts about one month, which

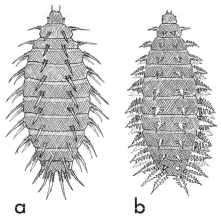

a   b

Figure 9.8 Final-instar larvae: (a) *Fannia canicularis* (lesser house-fly);
(b) *Fannia scalaris* (latrine-fly).

is considerably longer than in *Musca domestica*, but may under favourable
conditions be completed within 13–22 days. Although adult *Fannia* species,
especially *F. canicularis*, often enter houses they do not settle on people or
on their food as much as house-flies, but tend to hover around the head in
an annoying manner.

9.4.3   Medical importance
Many of the pathogens transmitted by house-flies are probably also spread
by *Fannia* species. They have been incriminated in cases of aural and urogenital myiasis, and larvae are sometimes found in stools, but as previously
stressed true intestinal myiasis does not occur in humans.

9.4.4   Control
The same control methods apply to species of *Fannia* as to *Musca*, but
particular attention should be given to the prevention and eradication of
Fannia scalaris breeding in latrines.

Further reading

Bidawid, S. P., Edeson, J. F. B., Ibrahim, J. and Matossian, R. R. (1978)
The role of non-biting flies in the transmission of enteric pathogens
(*Salmonella* species and *Shigella* species) in Beirut, Lebanon. *Annals of
Tropical Medicine and Parasitology*, **72**, 117–21.

Chavasse, D. C., Shier, R. P., Murphy, O. A., Huttly, S. R. A., Cousens, S. N.
and Akhtar, T. (1999) Impact of fly control on childhood diarrhoea in
Pakistan: community-randomised trial. *Lancet*, **353**, 22–5.

Cohen, D., Green, M., Block, C. *et al.* (1991) Reduction of transmission of
shigellosis by control of houseflies (*Musca domestica*). *Lancet*, **337**, 993–7.

Curtis, C. (1998) The medical importance of domestic flies and their control. *Africa Health*, **20** (6), 14–15.

Emerson, P. M., Bailey, R. L., Walraven, G. E. and Lindsay, S. W. (2001) Human and other faeces as breeding media of the trachoma vector *Musca sorbens*. *Medical and Veterinary Entomology*, **15**, 314–20.

Graczk, T. K., Knight, R., Gilman, R. H. and Cranfield, M. R. (2001) The role of non-biting flies in the epidemiology of human infectious diseases. *Microbes and Infection*, **3**, 231–5.

Greenberg, B. (1971) *Flies and Disease. Volume 1: Ecology, Classification and Biotic Associations*. Princeton, NJ: Princeton University Press.

Greenberg, B. (1973) *Flies and Disease. Volume 2: Biology and Disease Transmission*. Princeton, NJ: Princeton University Press.

Krafsur, E. S. and Moon, R. D. (1997) Bionomics of the face-fly, *Musca autumnalis*. *Annual Review of Entomology*, **42**, 503–23.

Levine, O. S. and Levine, M. M. (1991) Houseflies (*Musca domestica*) as mechanical vectors of shigellosis. *Review of Infectious Diseases*, **13**, 688–96.

Lindsay, D. R., Stewart, W. H. and Watt, J. (1953) Effect of fly control on diarrheal diseases in an area of moderate morbidity. *Public Health Reports, Washington DC*, **68**, 361–7.

Olsen, R. A. (1998) Regulatory action criteria for filth and other extraneous materials: III. Review of flies and foodborne enteric diseases. *Regulatory Toxicology and Pharmacology*, **28**, 199–211.

Rahuma, N., Ghenghesh, K. S., Ben Aissa, R. and Elamaari, A. (2005) Carriage by the housefly (*Musca domestica*) of multiple-antibiotic-resistant bacteria that are potentially pathogenic to humans, in hospitals and other urban environments in Misurata, Libya. *Annals of Tropical Medicine and Parasitology*, **99**, 795–802.

Skidmore, P. (1985) *The Biology of the Muscidae of the World*. Series Entomologica 29. Dordrecht: W. Junk.

Sukontason, K., Bunchoo, M., Khantawa, K., Sukontason, S., Piangjai, S. and Choochote, W. (2000) *Musca domestica* as a mechanical carrier of bacteria in Chiang Mai, North Thailand. *Journal of Vector Ecology*, **25**, 114–17.

West, L. S. (1951) *The Housefly: its Natural History, Medical Importance and Control*. Ithaca, NY: Comstock; Cornell University Press.

Zumpt, F. (1973) *The Stomoxyine Biting Flies of the World. Diptera: Muscidae. Taxonomy, Biology, Economic Importance and Control Measures*. Stuttgart: Gustav Fischer.

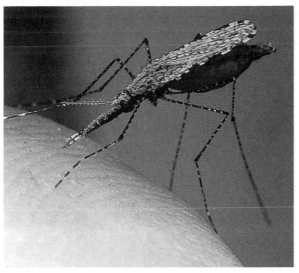

Actual length 6–7 mm

Plate 1 Adult female *Anopheles* (*An. farauti*) blood-feeding. Note head down and abdomen up, long palps and black and white blocks of scales on the wings. Not all *Anopheles* have white marks on their legs. (Courtesy of Desmond H. Foley.)

Actual length 7 mm

Plate 2 Adult female *Culex* (*Cx. quinquefasciatus*). Note body more or less parallel to surface (compare Plate 1), dull brown colour and lack of conspicuous ornamentation on the body or legs. (This photograph is taken from the Food and Environmental Hygiene Department's website and is reproduced under a licence from the Government of Hong Kong Special Administrative Region. All rights reserved.)

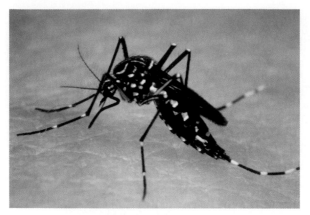

Actual length 7 mm

Plate 3 Adult female *Aedes* (*Ae. aegypti*) blood-feeding. Note body more or less parallel to surface (compare Plate 1) and body and legs covered with black and silvery scales. (Courtesy of Leonard E. Munstermann.)

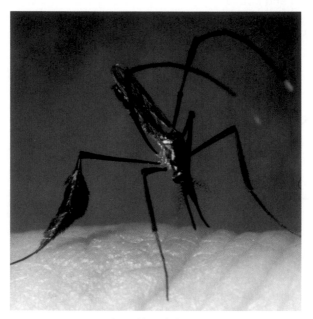

Actual length 5 mm

Plate 4 Adult female *Sabethes* (*Sa. cyaneus*). Found only in Central and South America. Note intense iridescent blue colour, hind legs curved over the head and middle legs with 'paddles'. (Courtesy of Leonard E. Munstermann.)

Actual length 3 mm

Plate 5 Adult female *Simulium* (*S. vittatum*). Note stout body, rather humped thorax, short antennae and clear colourless wings with faint wing veins (compare Plate 7). (Courtesy of Klaus Bolte and the Canadian Forest Service, Ottawa.)

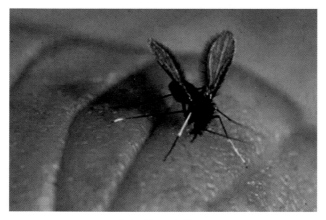

Actual length 2 mm

Plate 6 Adult female phlebotomine sand-fly (*Lutzomyia longipalpis*) blood-feeding. Note long thin legs and hairy wings held above the body. (Courtesy of Richard W. Ashford.)

Actual length 1–2 mm

Plate 7 Adult female *Culicoides* (*C. sonorensis*) blood-feeding. Note wings folded over abdomen. Wings lack scales but have dark and pale markings (compare Plate 5). (Courtesy of the Cooperative Extension Service, University of California, Davis, and Edward Schmidtmann.)

Actual length 8–10 mm

Plate 8 Adult female *Chrysops* (*C. pikei*). Note long antennae, dark band across each wing and iridescent eyes, which are widely spaced, so a female. (Courtesy of Sturgis McKeever, Georgia Southern University, USA, Insect Images.)

Actual length 8–11 mm

Plate 9 Adult tsetse-fly (*Glossina morsitans* group). Note proboscis projecting straight in front of head, wings folded scissor-like over the body, and 'hatchet cell' on wing. (Courtesy of John W. McGarry.)

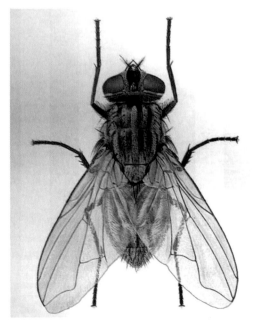

Actual length 6–9 mm

Plate 10 Adult house-fly (*Musca domestica*). Note four dark stripes on thorax and the tip of wing vein 4 almost touching vein 3. (Courtesy of the Armed Forces Pest Management Board (www.afpmb.org). Image provided by Ciba-Geigy Corporation.)

Actual length 13 mm

Plate 11 Larvae of the tumbu fly (*Cordylobia anthropophaga*). Found only in Africa. Note squat shape and body covered with very small spicules. (Courtesy of Audio Visual, LSHTM/Wellcome Photo Library.)

Actual length 15–17 mm

Plate 12 Larva of a New World screwworm fly (*Cochliomyia hominivo-rax*) (larvae of Old World screwworms (*Chrysomya* species) are almost identical). Note typical maggot shape and bands of minute spicules encircling body segments. (Courtesy of John W. McGarry.)

(a)

(b)

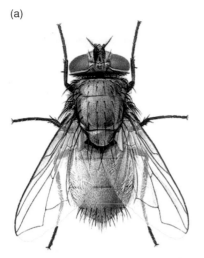

Actual length 10 mm

Plate 13 Adult greenbottle fly (*Lucilia sericata*). (a) Dorsal view and (b) lateral view. Note metallic green colour of body, numerous bristles on the thorax and abdomen, and uncoloured and transparent wings. (a) Courtesy of the Armed Forces Pest Management Board (www.afpmb.org); image provided by Ciba-Geigy Corporation. (b) Courtesy of Jay Dykes.

Actual length 12–14 mm

Plate 14 Adult bluebottle fly (*Calliphora vicinia*). Note metallic dark bluish body, numerous bristles on the thorax and abdomen, and uncoloured and transparent wings. (Courtesy of Jay Dykes.)

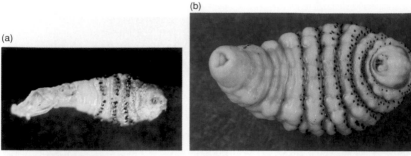

(a)

(b)

Actual length 6–10 mm

Actual length 14–18 mm

Plate 15 Human bot-fly (*Dermatobia hominis*). (a) second-instar larva and (b) third-instar larva. Note bands of coarse spines encircling anterior body segments. The bot-fly is restricted to Central and South America. (a) Courtesy of John W. McGarry. (b) Reprinted with permission from the *Annual Review of Entomology*, Volume 52; © 2007 by Annual Reviews (www.annualreviews.org). Photo by Craig Welch.

Actual length 2–4 mm

Plate 16 Adult cat flea (*Ctenocephalides felis*). Note brown body bearing long legs, pronotal comb of spines on first thoracic segment just behind the head, and a smaller genal comb under the head. (Courtesy of John W. McGarry.)

Actual length 3 mm

Plate 17 Adult head louse (*Pediculus capitis*). Note pale colour of body but dark colour of partially digested blood in the gut, and distinct head with short antennae. The body louse looks virtually the same. (Courtesy of Oxford Scientific.)

Actual length 5–6 mm

Plate 18 Adult bedbug (*Cimex lectularius*). Note dark brown colour, winged collar behind the head and long antennae and small bulging eyes. (Courtesy of Natural History Museum, London.)

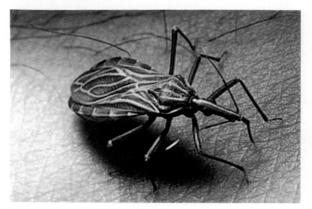

Actual length 32 mm

Plate 19 A typical adult triatomine bug (*Rhodnius prolixus*). Note long snout-like head bearing two long thin antennae, and wings folded over abdomen. (Courtesy of WHO/TDR, photo by Sinclair Stammers.)

Actual length 35–40 mm

Plate 20 A typical adult cockroach (*Periplaneta americana*). Note very long filamentous antennae, dark brown wings crossed over body concealing the abdomen, and spines and bristles on the legs. (Courtesy of Killgerm Group Ltd, UK.)

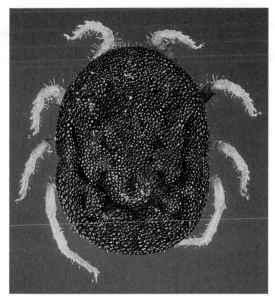

Actual length 8–10 mm

Plate 21 An adult soft (argasid) tick (*Ornithodorus savignyi*, but dorsally indistinguishable from. *O. moubta*). Note almost round shape of body, which is covered with fine tubercles, absence of a scutum, and capitulum invisible as hidden underneath the body. (Courtesy of Alan R. Walker and the University of Edinburgh.)

Actual length 8 mm

Plate 22 Adults of a hard (ixodid) tick (*Amblyomma variegatum*): female (left) with small scutum and festoons and male (right), which is smaller and has the scutum covering much of the body; festoons are present. Not all hard ticks have festoons. (Courtesy of Alan R. Walker and the University of Edinburgh.)

(a)

(b)

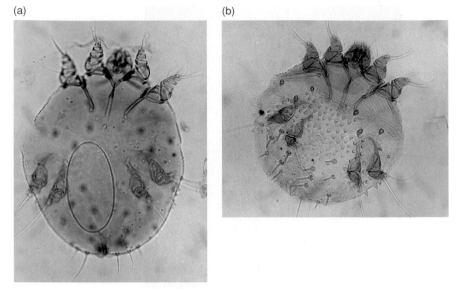

Actual length 0.30–0.45 mm

Plate 23 Adult female scabies mite (*Sarcoptes scabiei*); note almost round shape and eight very short and conical legs: (a) showing a large egg; (b) showing numerous peg-like spines, short setae and striations on the body. (Courtesy of John W. McGarry.)

(a)

(b)

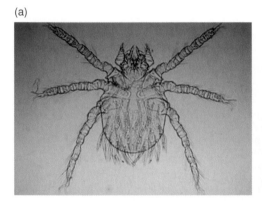

Actual length 0.25 mm

Plate 24 Larvae of scrub typhus mites (*Leptotrombidium* species): (a) ventral view of *L. orientale* showing plumose setae on the body and the three pairs of legs; (b) dorsal view showing large claws on legs, body setae and the large palps and mouthparts. (a) Courtesy of Tai Soon Yong, Web Atlas of Medical Parasitology of the Korean Society for Parasitologists. (b) Courtesy of John W. McGarry.

10

Flies and myiasis

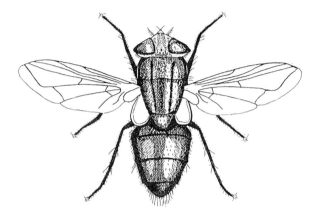

Myiasis is the invasion of organs and tissues of humans or other vertebrate animals by *fly* larvae, which at least for some time feed upon the living or dead tissues or, in the case of intestinal myiasis, on the host's ingested food.

10.1   Types of myiasis

Myiasis may be accidental, obligatory or facultative.

Accidental myiasis usually involves eating food that is contaminated by eggs or larvae of flies that are not parasitic in mammals, such as house-flies. Although the larvae may survive for some time in the intestine, no flies are specially adapted to cause intestinal myiasis in humans. (In contrast, obligatory intestinal myiasis occurs in animals.) The presence of larvae in the human intestine may nevertheless cause considerable discomfort, abdominal pain and diarrhoea, which may be accompanied by discharge of blood and vomiting. Living larvae may be passed in excreta or vomit.

In *obligatory* myiasis it is essential for the fly *maggots* (larvae) to live on a live host for at least a part of their life. For example, larvae of *Cordylobia anthropophaga*, *Cochliomyia hominivorax*, *Chrysomya bezziana*, *Dermatobia hominis* and *Wohlfahrtia magnifica* are obligatory parasites of humans and other vertebrates.

In contrast, in *facultative* myiasis larvae are normally free-living, often attacking carcasses, but under certain conditions may infect living hosts. Several types of fly, including species of *Calliphora*, *Lucilia* (= *Phaenicia*), *Phormia* and *Sarcophaga*, which normally breed in meat or carrion, may cause facultative cutaneous myiasis in people by infecting festering sores and wounds. Occasionally facultative urogenital myiasis occurs in humans, usually involving larvae of *Musca* or *Fannia*. Ovipositing flies attracted to unhygienic discharges lay their eggs near genital orifices, and on hatching the minute larvae enter the genital orifice and pass up the urogenital tract. Considerable pain may be caused by larvae obstructing these passages, and mucus, blood and eventually larvae may be discharged during urination.

Different terms are used to describe myiasis which affects different parts of the body – for example, *cutaneous*, dermal or subdermal myiasis; *urogenital* myiasis; *ophthalmic* or ocular myiasis; *nasopharyngeal* myiasis; and *intestinal*, gastrointestinal or enteric myiasis. When larvae burrow just under the surface layers of the skin this is sometimes called *creeping eruption* or creeping myiasis; when boil-like lesions are produced the term *furuncular* myiasis may be used; and when wounds become infested this is often referred to as *traumatic* myiasis.

When larvae occur in wounds, sores and dermal or subdermal tissues, their removal under aseptic conditions is usually relatively simple. When, however, they are more deeply imbedded in the underlying tissues, or when they have penetrated the mucous membranes, eyes, frontal sinuses or cavities, their removal is more difficult and surgery may be needed. Major and irreversible damage may have been done by the larvae.

The biologies and medical importance of the principal types of flies causing facultative and obligatory myiasis in humans are outlined below.

10.2   Classification

Flies described in this chapter are in three families: the Calliphoridae, Sarcophagidae and Oestridae. The family Calliphoridae can be conveniently divided into two groups. One group contains the *non-metallic* flies such as the Congo floor-maggot fly (*Auchmeromyia senegalensis*) and the tumbu fly (*Cordylobia anthropophaga*). The other group contains the *metallic* calliphorids, such as the blowflies, comprising the bluebottles (*Calliphora*) and the greenbottles (*Lucilia* (= *Phaenicia*)), and the New World screwworms (*Cochliomyia hominivorax*) and Old World screwworms (*Chrysomyia bezziana*).

10.3   Calliphoridae: non-metallic flies

10.3.1 *Cordylobia anthropophaga*
External morphology
Cordylobia anthropophaga, known as the tumbu or mango fly, is found in Africa from Ethiopia in the north through West and East Africa to Natal and the Transvaal in the south. It is not found outside Africa except possibly in south-western Saudi Arabia, and in 2006 a traveller returning to the UK from Portugal was infected with tumbu fly larvae.

Adults are robust, relatively large flies, 9–12 mm long, dull yellowish to light brown but with two dark grey poorly defined dorsal longitudinal thoracic stripes (Fig. 10.1). There are four visible abdominal segments which are more or less *equal* in length (compare *Auchmeromyia senegalensis*, in which the second abdominal segment is markedly *longer* than the others). The wings are slightly brownish.

Life cycle
Females lay 100–300 eggs in several batches on dry soil or sand in shady places, especially those contaminated with the urine or excreta of humans, rodents, dogs or monkeys. Females also oviposit on underclothes or soiled babies' nappies (diapers) placed on the ground. The white banana-shaped eggs hatch after 1–3 days. Larvae attach themselves either directly to a suitable host or to washed clothing placed on the ground to dry. Larvae get transferred to people if clothing is not ironed before it is worn. On a host a larva buries itself completely except for its posterior spiracles, situated at the tip of the abdomen, which remain in contact with the air. Newly hatched larvae can remain alive for 9–15 days on the ground in the absence of a suitable host.

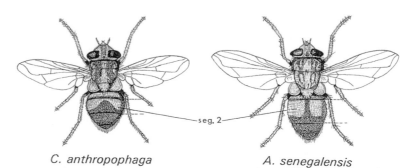

C. anthropophaga   A. senegalensis

Figure 10.1 Adults of the tumbu fly (*Cordylobia anthropophaga*) and the Congo floor-maggot fly (*Auchmeromyia senegalensis*), showing the longer second abdominal segment in *A. senegalensis*.

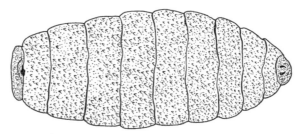

Figure 10.2 Final-instar larva of the tumbu fly (*Cordylobia anthropophaga*).

Minute first-instar larvae are maggot-shaped (like the larvae of house-flies), the second-instar larvae are club-shaped, and the third and final instar larvae are rather fat, broadly oval-shaped, yellowish-white maggots about 11–15 mm long. They are covered with numerous **spicules**, which are often, but not always, grouped into three or more transverse rows per segment (Fig. 10.2, Plate 11). After 8–12 days mature larvae wriggle out of boil-like swellings and fall to the ground, where they bury themselves and turn into puparia. Adult flies emerge 8–15 days later and feed on rotting fruit, carrion and faeces. Adults readily enter houses, where they may lay their eggs on mud floors, especially if children have urinated on them.

Medical importance
Larvae of *Cordylobia* cause boil-like (furuncular) swellings on almost any part of the body. Although these swellings may become sore and inflamed, and even quite hard and exude serous fluids, they do not usually contain pus. Generally only one or two larvae are found in a patient; however, more than 60 larvae have been recovered from one person, mostly in separate

lesions. A larva can be removed by covering the hole in the swelling with medicinal liquid paraffin, which prevents it from breathing through its posterior spiracles. Consequently a larva wriggles a little further out of the swelling to protrude its spiracles, and this lubricates the pocket in the skin. The larva can then be extracted by gently pressing around the swelling.

Infections can be prevented, or at least minimized, by ensuring that clothes, bed linen and towels are not spread on the ground to dry, and also by ironing clothing.

Dogs and rats are commonly infected with tumbu larvae.

10.3.2 *Auchmeromyia senegalensis*
External morphology
Auchmeromyia senegalensis, commonly known as the Congo floor-maggot fly, although not strictly speaking causing myiasis, is described here because adults are often confused with those of *Cordylobia anthropophaga*. The species occurs throughout Africa south of the Sahara and also in the Cape Verde Islands.

Adults are very similar to *C. anthropophaga* but are distinguished by the second abdominal segment being about *twice as long* as any of the others (Fig. 10.1), whereas in the tumbu fly all segments are about equal in length.

Life cycle and medical importance
Some 300 eggs are laid in batches of about 50 on the dry sandy floors of mud huts. They hatch after 1–3 days and the larvae hide in cracks and crevices in the floor, especially under beds and sleeping-mats. At night they crawl out and take blood-meals from people sleeping on the floor, after which the now pinkish larvae return to their hiding places. Larvae may feed four or five times a week, but can withstand long periods of starvation in the absence of suitable hosts. There are three larval instars, each requiring at least two blood-meals. Under optimum conditions larval development is completed within 3–4 weeks, but it may be prolonged up to three months if larvae fail to obtain regular feeds. Third-instar larvae, unlike those of *Cordylobia anthropophaga*, are not covered with conspicuous spicules (Fig. 10.3). Because larvae cannot climb, people will not be attacked if they sleep on beds raised from the floor. Mature larvae pupate in cracks or directly on the surface of mud floors. Adults emerge from the puparia after about 9–16 days and, like tumbu flies, feed on rotting fruit and faeces.

The Congo floor-maggot fly was formerly common in certain parts of Africa, but due to changes in lifestyle it is becoming increasingly rare, and these days is not much of a problem.

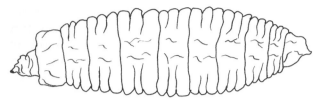

Figure 10.3 Final-instar larva of the Congo floor-maggot fly (*Auchmeromyia senegalensis*).

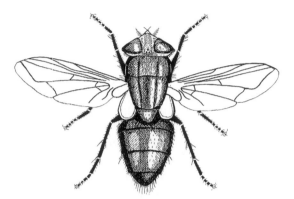

Figure 10.4 Adult of the New World screwworm fly (*Cochliomyia hominivorax*).

10.4   Calliphoridae: metallic flies

10.4.1 New World screwworms

New World screwworms (*Cochliomyia hominivorax*) formerly occurred in the southern states of the USA, through Mexico, Central America, the Caribbean islands, into the northern parts of South America. However, following successful eradication programmes, achieved by the release of thousands of sterilized male flies, the screwworm has been eradicated from the USA, Mexico, the Virgin islands, Curacao and Central America except for parts of Panama. An ongoing programme in Jamaica has almost eradicated the fly.

In 1988 a population of the New World screwworm was discovered in Libya, but it was eradicated in 1991.

External morphology

Adult flies of *C. hominivorax* are 8–10 mm long, metallic green to bluish green, and have three distinct dark longitudinal stripes on the dorsal surface of the thorax (Fig. 10.4). **Dorsal bristles** on the thorax, like those of *Chrysomya*, are poorly developed, thus distinguishing screwworm flies from *Lucilia* (= *Phaenicia*) and *Calliphora*, which have well-developed

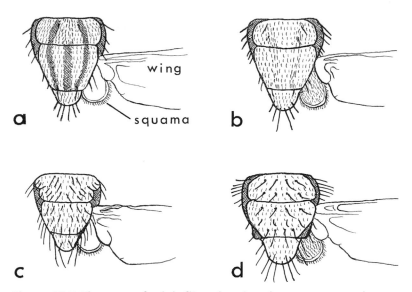

Figure 10.5 Thoraces of adult flies, showing the presence or absence of prominent dorsal bristles; and bases of the right wings, showing the presence or absence of fine hairs on the squama: (a) *Cochliomyia*: note the three dark thoracic stripes and squama lacking hairs; (b) *Chrysomya*: note the absence of prominent thoracic stripes and the hairy squama; (c) *Lucilia*: note the prominent thoracic bristles and squama lacking hairs; (d) *Calliphora*: note the prominent thoracic bristles and hairy squama.

bristles. The *squama*, a membranous lobe on the posterior border of the wing near the thorax (Fig. 10.5a), is *devoid* of hairs, as in *Lucilia*.

Life cycle

Females lay batches of 10–400 eggs on the edges of wounds, scabs, sores or scratches, also on dried blood clots and on diseased or healthy mucous membranes such as the nasal passages, eyes, ears, mouth and vagina. In newborn babies eggs may be laid in the umbilicus. Eggs hatch after 11–24 hours and the active larvae burrow deeply into living tissues and feed gregariously. There are three larval instars and the third-instar, which is formed after 2–3 days, is about 15–17 mm long and typically *maggot*-shaped. They are distinguished from house-fly maggots by the presence of distinct **bands of spicules** encircling the anterior margins of all body segments (Fig. 10.6a, Plate 12), and by the dark part of the tracheal trunks from the posterior spiracles extending into 3–4 abdominal segments. Larvae tend to penetrate deeply into tissues, so that infections near the eyes, nose and mouth can cause considerable destruction, often accompanied by putrid-smelling discharges and ulcerations.

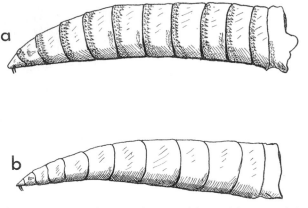

Figure 10.6 Final-instar larvae: (a) *Cochliomyia* (*Chrysomya* larvae are similar) showing bands of spicules on the segments; (b) *Lucilia* (*Calliphora* larvae are almost identical) lacking prominent spicules.

After 4–12 days the larvae reach maturity and wriggle out of the wounds or passages they have excavated and drop to the ground, where they bury in the soil and pupate. In warm weather the puparial stage lasts about 7–10 days, but in cooler weather it may be prolonged for many weeks or even months. The life cycle from egg to adult is usually about 2–3 weeks.

10.4.2 Old World screwworms

The genus *Chrysomya* contains many species and is common in the tropics. About 10 species are known to cause myiasis in humans, but only *Chrysomya bezziana* is important because its larvae, like those of *Cochliomyia hominivorax*, are **obligatory** parasites of living tissues. *Chrysomya bezziana* occurs throughout tropical Africa, the Indian subcontinent, in most of South-east Asia to China, and the Philippines to Papua New Guinea; it also has been introduced to several countries on the west coast of the Persian Gulf, including Iraq, Iran and Saudi Arabia. Larvae of other *Chrysomya* species are not obligatory parasites, and often develop in carrion and decomposing matter. However, *facultative* myiasis-producing species such as *Chrysomya albiceps* and *C. megacephala* of the Old World have invaded Central and South America.

External morphology

Adult *Chrysomya bezziana* are similar to *Cochliomyia hominivorax*, but they lack the distinctive longitudinal thoracic stripes (Fig. 10.5b), and the dorsal surface of the **squama** is covered with fine hairs (Fig. 10.5b). Larvae are very similar to those of *C. hominivorax*, but usually have 4–6 finger-like processes on the anterior spiracles, not the usual 7–9 found in *C. hominivorax* and by the dark part the tracheal trunks from the posterior spiracles extending into only one abdominal segment. In practice separation is usually easy: screwworms of the Americas belong to the genus *Cochliomyia* whereas those of the Old World belong to the genus *Chrysomya*.

Life cycle

The life cycle of *Chrysomya bezziana* is very similar to that of *Cochliomyia hominivorax*. About 150–500 eggs are deposited in wounds, open sores, scabs, ulcers, scratches or on mucous membranes, especially those contaminated with discharges. They hatch after 24 hours or less and newly emerged larvae burrow through the skin to the underlying tissues, where they commonly remain congregated together. Larvae complete their development in 5–6 days and then wriggle out of the wounds and drop to the ground, where they bury themselves and pupate. The puparial period lasts about 7–9 days in warm weather, but is prolonged to several weeks or even months during cold weather. The life cycle from egg to adult usually takes about 2–3 weeks.

Adults of both New and Old World screwworms frequently feed on decomposing corpses, decaying matter, excreta and flowers.

Medical importance of screwworms

Larvae of both *Chrysomya bezziana* and *Cochliomyia hominivorax* cause **obligatory** myiasis in humans, resulting in considerable damage and disfigurement, especially if the face is involved. When larvae invade natural orifices, such as the nose, mouth, eyes or vagina, they can cause excruciating pain and misery. In one patient suffering from a nasal infection 385 larvae of *C. hominivorax* were removed over a 9-day period! Larvae of both species may eat their way through the palate and as a result impair speech.

Chrysomya bezziana causes more cases of myiasis in people living in India and other parts of Asia than it does in Africa. *Chrysomya* adults often visit faeces, so are potential vectors of several pathogens.

Myiasis cases should be treated immediately because the very rapid larval development can cause permanent damage. Maggots in open wounds or body openings can be removed by irrigating infested areas with ethanol or chloriform mixed with vegetable oil. Recently oral or topical use of ivermectin has resulted in killing the larvae. Surgery may be necessary to expose deeply embedded larvae. There are serious problems with ocular myiasis and delicate surgery may be needed to remove the larvae; permanent damage may have been done.

Both screwworm species, but particularly *C. hominivorax*, cause myiasis in cattle, goats, sheep and horses and cause enormous economic losses to the livestock industry. The technique of releasing millions of laboratory-reared sterile male flies into areas infested with screwworms has given good control, and has even resulted in the eradication of *C. hominivorax* from the USA, Mexico, Guatemala, Belize, Honduras,

El Salvador, Nicaragua, Costa Rica and some Caribbean islands, as well as from Libya.

10.4.3 Greenbottles (*Lucilia = Phaenicia*)
External morphology
Several species of greenbottles occur in the genus *Lucilia*, and although the genus has a worldwide distribution most species are in northern temperate regions. Some entomologists use the generic name *Phaenicia* instead of *Lucilia*.

Adults are mostly metallic or coppery green (Plate 13), usually a little smaller (about 10 mm long) and a little less bristly than species of *Calliphora* (bluebottles). As in *Calliphora*, prominent **bristles** occur on the dorsal surface of the thorax (Fig. 10.5c), but the **squama** of the wing lacks hairs (Fig. 10.5c), whereas in bluebottles it is hairy dorsally. *Lucilia sericata* is the commonest species, occurring in the Americas, Europe, Asia, Africa, Australia and in most other areas of the world. Another common species, *L. cuprina*, occurs mainly in Africa, Asia and Australia.

Life cycle
Greenbottles usually lay their eggs on meat, fish, carrion and decomposing carcasses, but occasionally on or near festering and foul-smelling wounds of humans and animals, as well as on excreta and decaying vegetable matter. Eggs hatch within 8–12 hours. Larvae are typically **maggot**-shaped (Fig. 10.6b), and they may have very small spicules but not conspicuous ones like the larvae of screwworms, which they otherwise tend to resemble. They can also be distinguished from screwworms by the posterior spiracles having a complete **peritreme**; in screwworms it is incomplete.

The larval period lasts about 4–8 days. Mature larvae bury in loose soil and pupate; the puparial period lasts about 6–14 days.

Adult flies frequently visit carrion, excreta, general refuse, decaying material, sores and wounds. They are particularly common around unhygienic situations where meat or dead animals are present. They are frequently abundant near slaughterhouses and piggeries. They commonly fly into houses, where they are particularly troublesome because of their noisy buzzing flight. The most common species infesting wounds of humans are *Lucilia sericata* and *L. cuprina*.

10.4.4 Bluebottles (*Calliphora*)
External morphology
There are numerous species of flies, known as bluebottles, belonging to the genus *Calliphora*. Although *Calliphora* has a worldwide distribution, bluebottles are more common in the northern temperate regions than in the tropical or southern temperate regions. The most important species with regard to myiasis are *Calliphora vicinia* and *C. vomitoria*.

Adults are robust flies, 8–14 mm long, dull metallic-bluish or bluish-black (Plate 14). As in *Lucilia* there are well-developed **bristles** on the thorax (Fig. 10.5d), but the **squama** of the wing is hairy on the dorsal surface (Fig. 10.5d) whereas in *Lucilia* it lacks hairs. The abdomen is rather more shiny than the thorax.

Life cycle
Bluebottle larvae look very similar to those of *Lucilia*, and the life cycle is also very similar to that described for *Lucilia*.

Medical importance of greenbottles and bluebottles
The dirty habit of blowflies (greenbottles and bluebottles) of feeding on excreta, decaying material and virtually all foods makes them potential vectors of numerous pathogens. However, their medical importance is usually associated with *facultative* myiasis.

Larvae of both *Lucilia* and *Calliphora* often develop in foul-smelling wounds and ulcerations, especially those producing pus. In hospitals they may be found underneath bandages and dressings of patients, especially when these have become contaminated with blood and pus. Such infections do not usually cause any serious harm since the larvae feed mainly on pus and dead tissues. In 1998 a patient in a hospital in the USA had about 100 larvae of *L. sericata* removed from nasal passages both manually and by nasal tracheal suction. Removal of maggots of *Lucilia* and *Calliphora* usually presents no problems because they can be picked out of wounds with sterile forceps and antibiotic dressings applied. Only very rarely do maggots invade healthy tissues. Occasionally intestinal myiasis is reported. This is usually caused by eating food contaminated with larvae of *Lucilia* or *Calliphora*, but usually the larvae are killed within the human alimentary canal and no serious harm is done. As previously emphasized (p. 152), there is no obligatory intestinal myiasis in humans.

Maggot therapy
Mayans of Central America, Australian aborigines and people in northern Myanmar have used maggot therapy for many centuries, and this practice continued in some parts of the world until about the 1950s. However, it seems that during the American Civil War (1861–5) the Confederate army surgeon J. F. Zacharias was the first Western physician to introduce maggots to clean wounds. Recently maggot therapy has been rediscovered and doctors, particularly in the USA and Britain but also in some other European countries, are advocating the use of sterile larvae to clean up leg ulcers, pressure sores and infected surgical wounds. Fly maggots most frequently used are those of *Lucilia sericata*. Some patients need considerable persuasion to accept this unusual form of treatment. By 2005 some twenty countries were using maggot therapy.

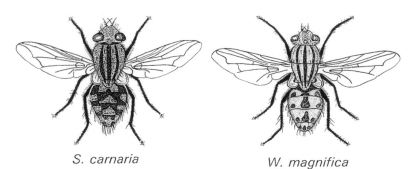

S. carnaria   *W. magnifica*

Figure 10.7 Adults of the flesh-flies *Sarcophaga carnaria* and *Wohlfahrtia magnifica*, showing differences in their abdominal markings.

Control of greenbottles and bluebottles

The principal breeding places of greenbottles and bluebottles are domestic refuse, rubbish tips, dustbins, offal and other wastes from slaughterhouses and meat packing factories. Larvae also occur in foods such as meat and fish left out in the sun to dry. Any methods reducing breeding sites are appropriate. Dustbins and garbage cans should have tight-fitting lids and be emptied once or twice a week. The outside of such bins and both sides of the lid can be sprayed with organophosphate or pyrethroid insecticides every 7–10 days, and adjacent walls and fences sprayed every two weeks.

Many of the insecticidal control measures used against house-flies (Chapter 9) are applicable to controlling blowflies.

10.5   Sarcophagidae: flesh-flies

Only species of *Sarcophaga* and *Wohlfahrtia* in the family Sarcophagidae are of medical importance. They are sometimes called flesh-flies. They cause myiasis, and possibly act as mechanical vectors of pathogens. Females are unusual because they are *larviparous*, that is they deposit first-instar larvae instead of laying eggs. They have a worldwide distribution.

10.5.1 *Sarcophaga*
External morphology

Sarcophaga species are large and hairy *non-metallic* flies, about 10–15 mm long. They are usually greyish and have three prominent black longitudinal stripes on the thorax. The abdomen is sometimes distinctly, but other times indistinctly, marked with squarish dark patches on a grey background giving it a *chequer-board* (chess-board) appearance (Fig. 10.7).

Life cycles and medical importance

Adult *Sarcophaga* do not lay eggs but deposit first-instar larvae, as do tsetse-flies and *Wohlfahrtia* species (i.e. they are *larviparous*). Larvae are

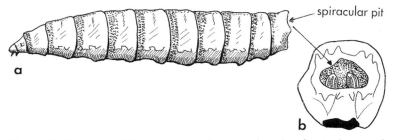

Figure 10.8 Larva of *Sarcophaga*, showing bands of spicules on the segments and the deep pit in which the posterior spiracles are located (*Wohlfahrtia* larvae are almost identical).

deposited in batches of 40–60, on decaying carcasses, rotting food and human and animal excreta, but sometimes in wounds. They are primarily scavengers. Larvae are *maggot*-shaped. They are distinguished from larvae of the Calliphoridae by the *posterior spiracles* being situated in a deep pit (Fig. 10.8) and thus difficult to see, and also by having bands of *spicules* on the body. Larvae of *Sarcophaga* are virtually indistinguishable from those of *Wohlfahrtia* species.

Larval development in hot weather lasts only 3–4 days, after which the larvae bury in the soil and pupate. The puparial stage lasts about 7–12 days.

Although larvae are usually deposited in carrion they very occasionally occur in wounds and cause *facultative* myiasis, but usually causing little damage because they feed mainly on necrotic tissues. They have more commonly been incriminated in accidental intestinal myiasis, causing considerable discomfort and pain before the larvae are expelled in the faeces. The most common species is *Sarcophaga cruentata* (= *haemorrhoidalis*), which is widely distributed in the Americas, Europe, Africa and Asia, while *S. carnaria* is common in the Palaearctic region. Because adults frequent festering wounds, excreta and decaying animal matter they may be mechanical vectors of various pathogens.

10.5.2 *Wohlfahrtia*
External morphology
Wohlfahrtia species are hairy flies, 8–15 mm long, about as large as bluebottles or a little larger. They are greyish and like *Sarcophaga* have three distinct black lines on the thorax. The dark markings on the abdomen, however, are not in the form of a chess-board pattern as in *Sarcophaga* species, but are usually present as *roundish* lateral spots and triangular-shaped dark markings along the midline (Fig. 10.7). There is, though, considerable variation and sometimes the dark marks are so large as to be more or less confluent, making the abdomen appear mainly black.

Life cycle and medical importance

As with *Sarcophaga* and tsetse-flies, adults of *Wohlfahrtia* deposit *larvae* not eggs. The most important species, *W. magnifica*, causes *obligatory* myiasis in humans and animals (e.g. camels, domestic livestock and dogs) in western Europe, the Middle East, North Africa and Central Asia to China. Some 120–170 larvae are deposited, often in several batches, in scratches, wounds, sores and ulcerations. In people the ears, eyes and nose are frequently infested, and this can cause deafness, blindness and even death. The North American species, *W. vigil*, will also deposit its larvae on unbroken skin if it is soft and tender, such as in babies and very young children, who are, therefore, more commonly attacked than adults. The maggots cause *furuncular* myiasis. *Wohlfahrtia* larvae have bands of *spicules*, and the *spiracles* situated in a deep pit (Fig. 10.8), closely resembling larvae of *Sarcophaga*. Larval development takes 5–9 days, after which mature larvae drop to the ground, bury themselves in loose soil and pupate. Adults emerge from the puparia after 8–12 days.

Like the larvae of screwworms, *Wohlfahrtia* larvae can burrow deeply into the tissues and cause considerable damage.

10.6   Oestridae: bot-flies

The family Oestridae comprises four subfamilies, three of which (Oestrinae, Gasterophilinae, Hypodermatinae) contain important obligate parasites of domesticated animals. Subfamily Cuterebrinae contains 57 species in two genera that cause myiasis in rodents, monkeys and livestock, and the human bot-fly (*Dermatobia hominis*). This fly causes *obligatory* myiasis in people and animals, mainly cattle, living in southern Mexico to northern Argentina.

10.6.1 *Dermatobia hominis*

External morphology

Adult *Dermatobia hominis* are a little larger (12–18 mm) than bluebottles (*Calliphora*), with a similar dark-blue *metallic-coloured* abdomen and dark bluish-grey thorax, but the head is mainly yellowish (Fig. 10.9). The mouthparts are vestigial. They are sometimes called the human bot-fly, or in the Americas ver macaque or tórsalo.

Life cycle

Dermatobia hominis occurs primarily in lowland forests, being especially common along woodland paths and at the margins of forest and scrub areas. These flies have an interesting and remarkable life history. Females glue 6–30 *eggs* to the body of other arthropods, such as day-flying mosquitoes (especially of the genus *Psorophora*), house-flies, stable-flies and occasionally even ticks.

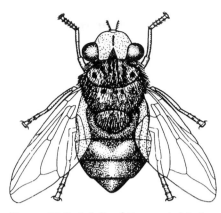

Figure 10.9 Adult of *Dermatobia hominis* (human bot-fly).

Embryos within the attached eggs mature into first-instar larvae after 4–7 days. They do not hatch, however, until the insect carrying the eggs settles on a human or some other warm-blooded animal to take a blood-meal or, as in the case of house-flies, to feed on sweat. Larvae then emerge from the eggs and drop onto the host's skin. Here, within 5–10 minutes, the larvae either enter the skin through the bite puncture or penetrate soft unbroken skin and burrow into the subcutaneous tissues. Each larva produces a boil-like (*furuncular*) swelling which has an opening through which the larva breathes. Occasionally two or three larvae may be found within a single swelling.

First-instar larvae are 1–1.5 mm long, more or less cylindrical in shape (Fig. 10.10a), and have the anterior half of the body covered with numerous *spines* of two different sizes. Second-instar larvae are a completely different shape, being enlarged anteriorly but with the posterior half of the body distinctly narrow, giving the appearance of a bottle with a long neck. Relatively large thorn-like *spines* encircle the middle segments (Fig. 10.10b). The third and final instar larvae are about 18–25 mm long, more or less oval in outline, and have relatively small *spines* on the anterior segments (Fig. 10.10c, Plate 15). A pair of distinct *flower-like* spiracles is present anteriorly.

Larval development is completed in a small pocket excavated in the subdermal tissues of the host, and lasts about 4–10 weeks. Mature larvae wriggle out of the skin and drop to the ground, where they pupate just under the surface of the soil. Adult flies emerge from puparia after about 4–11 weeks, but are rarely seen.

Medical importance

Larvae of *Dermatobia hominis* invade the subcutaneous tissues of humans on various parts of the body, including the head, arms, abdomen,

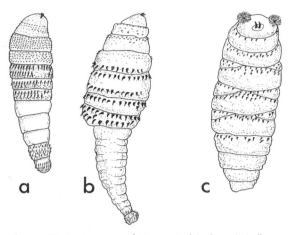

Figure 10.10 Larvae of *Dermatobia hominis* (human bot-fly), showing small and larger spines: (a) first instar; (b) second instar; (c) third and final instar.

buttocks, thighs, scrotum and axillae. They produce boil-like (*furuncular*) swellings which suppurate and this may attract other myiasis-producing flies. They can cause considerable discomfort and pain. The *long duration* of the larval life (up to 10 weeks or more) means that infected persons may be encountered in almost any part of the world following air travel.

Because of their numerous spines and shape larvae are often difficult to remove by squeezing them out; consequently surgical removal under sterile conditions may be necessary, frequently accompanied by a local anaesthetic.

10.7   Other myiasis-producing flies

The black blowfly (*Phormia regina*), found in North America, Europe and northern parts of Asia, breeds mainly in carrion but sometimes causes facultative myiasis in humans. *Cochliomyia macellaria* is found from North to South America and develops primarily on carrion, but occasionally larvae are found in wounds, and so cause facultative myiasis. It is of only minor medical importance.

Several species of flies cause myiasis in livestock, such as sheep and goats (e.g. *Oestrus ovis*), cattle (e.g. *Hypoderma bovis*) and horses and donkeys (e.g. *Gasterophilus haemorrhoidalis*). Occasionally humans become infected with their maggots. Those most likely to suffer are people working with the flies' natural hosts: for example, shepherds looking after sheep may become infected with *O. ovis*.

Further reading

Abram, L. J. and Froimson, A. I. (1987) Myiasis (maggot infection) as a complication of fracture management: a case report and review of the literature. *Orthopedics*, **10**, 625–7.

Arbit, E., Varon, R. E. and Brem, S. S. (1986) Myiatic scalp and skull infection with Diptera *Sarcophaga*: a case report. *Neurosurgery*, **18**, 361–2.

Catts, E. P. (1982) Biology of New World bot flies: Cuterebridae. *Annual Review of Entomology*, **27**, 313–38.

Colwell, D. D., Hall, M. J. R. and Scholl, P. J. (eds.) (2006) *The Oestrid Flies: Biology, Host–Parasite Relationships, Impact and Management.* Wallingford: CABI.

Gabaj, M. M., Gusbi, A. M. and Awan, M. A. Q. (1989) First human infestations in Africa with larvae of the American screw-worm, *Cochliomyia hominivorax* Coq. *Annals of Tropical Medicine and Parasitology*, **83**, 553–4.

Graham, O. H. (ed.) (1985) Symposium on eradication of the screw-worm from the United States and Mexico. *Miscellaneous Publications of the Entomological Society of America*, **62**, 1–68.

Greenberg, B. and Kunich, J. C. (2002) *Entomology and the Law: Flies as Forensic Indicators.* Cambridge: Cambridge University Press.

Hall, M. J. R. and Wall, R. (1995) Myiasis of humans and domestic animals. *Advances in Parasitology*, **35**, 257–334.

Kersten, R., Shoukrey, N. M. and Tabbara, K. F. (1986) Orbital myiasis. *Ophthalmology*, **93**, 1228–32.

Krafsur, E. S., Whitten, C. J. and Novy, J. E. (1987) Screw-worm eradication in North and Central America. *Parasitology Today*, **3**, 131–7.

Lane, R. P., Lovell, C. R., Griffiths, W. A. D. and Sonnex, T. S. (1987) Human cutaneous myiasis: a review and report of three cases due to *Dermatobia hominis*. *Clinical and Experimental Dermatology*, **12**, 40–5.

Nunzi, E., Rongioletti, F. and Rebora, A. (1986) Removal of *Dermatobia hominis* larvae. *Archives of Dermatology*, **122**, 140.

Sherman, R. A., Hall, M. J. R. and Thomas, S. (2000) Medicinal maggots: an ancient remedy for some contemporary afflictions. *Annual Review of Entomology*, **45**, 55–61.

Shoaib, K. A. A., McCall, P. J., Goyal, R., Loganathan, S. and Richmond, W. D. (2000) First urological presentation of New World screw worm (*Cochliomyia hominovorax*) myiasis in the United Kingdom: a case report. *British Journal of Urology – International*, **86**, 16–17.

Smith, K. G. V. (1986) *A Manual of Forensic Entomology.* London: British Museum (Natural History); Ithaca, NY: Cornell University Press, pp. 93–137.

Spradbery, J. P. (1991) *A Manual for the Diagnosis of Screw-Worm Fly.* Canberra: Commonwealth Scientific and Industrial Research Organisation, Division of Entomology.

Zumpt, F. (1965) *Myiasis in Man and Animals in the Old World: a Textbook for Physicians, Veterinarians and Zoologists.* London: Butterworth.

11

Fleas (Siphonaptera)

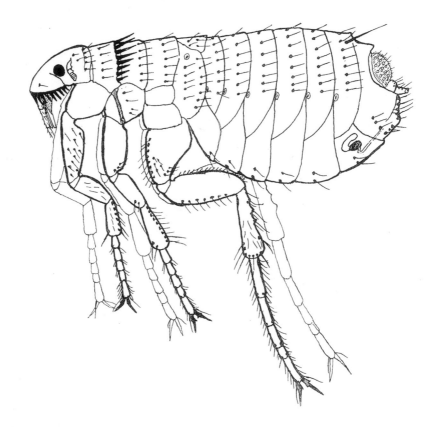

There are more than 2200 species and subspecies of fleas in about 220 genera, but only relatively few are important pests of humans. About 94% of species bite mammals while the remainder are parasitic on birds. Fleas occur almost worldwide, but many have a more restricted distribution; for example the genus *Xenopsylla*, which contains important plague vectors, is confined to the tropics and warmer parts of some temperate countries.

Medically the most important fleas are *Xenopsylla* species, such as *X. cheopis*, which is a vector of plague (*Yersinia pestis*) and flea-borne murine typhus (*Rickettsia typhi*). Fleas in the genus *Ctenocephalides* may be intermediate hosts of cestodes (*Dipylidium caninum*, *Hymenolepis diminuta*). Fleas may also be vectors of tularaemia (*Francisella tularensis*), and the chigoe or jigger flea (*Tunga penetrans*) 'burrows' into people's feet.

11.1   External morphology

Adult fleas are relatively small (1–6 mm), more or less oval insects, *compressed* laterally and varying from light to dark brown (Plate 16). Wings are absent, but there are three pairs of powerful legs, the hind pair of which are specialized for jumping. The legs, and much of the body, are covered with bristles and small spines.

The head is approximately triangular, bears a pair of conspicuous eyes (a few species are eyeless), and short three-segmented more or less club-shaped *antennae* which lie in depressions behind the eyes. The mouthparts point downwards. In some species a row of coarse, well-developed tooth-like spines, collectively known as the *genal comb* or genal ctenidium, is present along the bottom margin of the head (Figs. 11.1, 11.2).

The thorax has three distinct segments: the pro-, meso- and metathorax. The posterior margin of the pronotum (i.e. dorsal part of the prothorax) may have a row of tooth-like spines forming the *pronotal comb* or pronotal ctenidium (Fig. 11.1). Fleas in some genera lack both pronotal and genal combs and are referred to as *combless* fleas (Fig. 11.2). In some genera fleas have both combs, while in other species the pronotal comb is present and the genal comb absent, but never the reverse (Fig. 11.2). A sternite called the mesopleuron is located above the middle pair of legs. In several genera, including *Xenopsylla*, which contains important plague vectors, this sternite is clearly divided into two parts by a thick vertical rod-like structure called the *meral rod*, pleural rod, or mesopleural suture or rod. The presence of this rod, combined with the absence of both genal and pronotal combs, indicates the genus *Xenopsylla* (Fig. 11.2). However, it must be stressed that the presence of a meral rod by itself does not identify fleas as *Xenopsylla* species, because fleas in several other genera have combs and a meral rod.

In female fleas the tip of the abdomen is more rounded than in males and is not upturned. Internally in about the sixth to eighth abdominal segments are one or two distinct brownish spermathecae (Fig. 11.1). However,

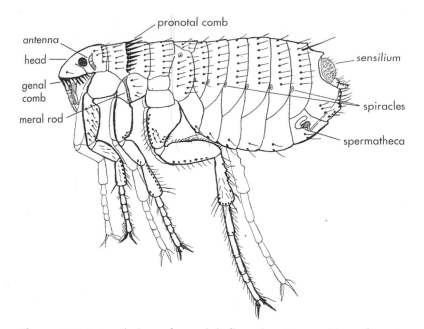

Figure 11.1 Lateral view of an adult flea, showing position of combs and the meral rod (pleural rod).

epidemiologically it is not important to separate the sexes because *both* take blood-meals and can be vectors. A sensory dome-shaped structure having setae, called the sensilium, is present dorsally on apparent segment 8 and aids fleas in detecting vibrations and temperature changes, and in host detection.

11.1.1 The alimentary tract of adult fleas

To understand the role fleas have in transmitting plague it is necessary to describe the alimentary tract and the method of blood-feeding.

Saliva, which contains anticoagulants, is injected into the host during feeding. Blood is sucked up through the pharynx and oesophagus into the bulbous *proventriculus* (Fig. 11.3), which is provided internally with numerous (250–450) backwardly projecting spines. It was previously accepted that these spines prevented the regurgitation of the blood-meal into the oesophagus, but this has not actually been proven. The proventriculus is nevertheless important in the mechanism of plague transmission.

Finally, the blood-meal enters a relatively large stomach (mid-gut), where it is digested. The hind-gut is continuous with a small dilated rectum, which has rectal glands that extract water so that the faeces pass out in a dry state.

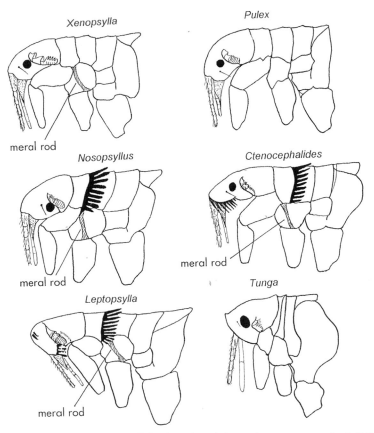

Figure 11.2 Diagrams of the head and thoracic segments of adult fleas, showing how the presence or absence of combs and the meral rod can distinguish six medically important genera.

11.2   Life cycle

Both sexes take blood-meals and are therefore equally important as disease vectors. The following account is a generalized description of the life cycle of fleas that feed on humans or animals, such as dogs, cats and commensal rats. The life cycle of the chigoe flea (*Tunga penetrans*) is described separately.

A female rodent flea leaves the host and deposits her eggs in debris which accumulates in the host's dwelling place, such as rodent burrows. Fleas that bite humans or their domestic pets, such as cat fleas, lay their eggs while they are still on the host and because they are not sticky they soon fall off the host and are mainly found in areas where hosts, such as cats or dogs, spend the most time. Eggs are very small (0.1–0.5 mm), oval, white or yellowish and lack any visible pattern. Adults commonly live for 10 days to 6 weeks, but sometimes for 6–12 months or even longer. During

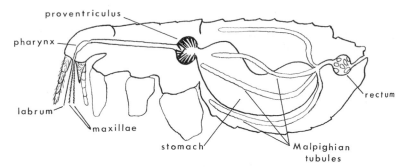

Figure 11.3 Diagrammatic representation of the alimentary canal of an adult flea, showing backwardly projecting spines in the proventriculus.

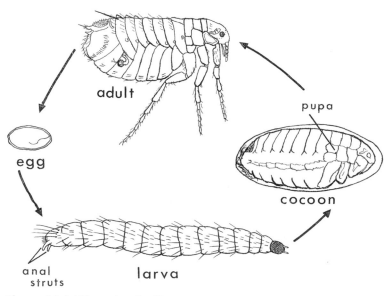

Figure 11.4 Life cycle of a flea.

her lifetime a female may lay 300–1000 eggs, mostly in batches of 3–25 a day.

Eggs usually hatch after 2–5 days but this depends on the species of flea, temperature and humidity. A minute legless larva emerges from the egg (Fig. 11.4). It has a small brownish head with a pair of very small antennae, followed by 13 pale brown, distinct and more or less similar segments. Each *segment* has a circle of setae near the posterior border. The last segment ends in a pair of finger-like ventral processes termed *anal struts*. The presence of these struts and setae on the body distinguish larval fleas from other insects of medical importance.

Larvae are very active. They avoid light, and shelter in cracks and crevices and amongst debris on floors of houses, or in nests or animal burrows. Occasionally, however, larvae are found on people who wear dirt-laden clothes, and sometimes in beds. Larvae feed on almost any organic debris but to successfully achieve adulthood it seems that larvae of many species must consume partly digested blood evacuated from the alimentary canal of adult fleas (i.e. adult flea faeces). There are usually three larval instars, but in a few species there are only two instars (e.g. *Tunga penetrans*, p. 178). The larval period is commonly 2–3 weeks, but varies according to species, and may be prolonged to 200 days or more by unfavourable conditions such as limited food supply and low temperatures. Mature larvae are 4–10 mm long. Unlike adult fleas, larvae die if relative humidities are either too low or too high.

At the end of the larval period the larva spins a whitish *cocoon* from silk produced by its salivary glands. Being sticky it becomes covered with fine particles of dust, debris and sand picked up from the floor of the host's home. Consequently cocoons are difficult to distinguish from their surroundings. About 2–3 days after having spun a cocoon around itself the larva pupates within the cocoon. Adults are ready to emerge from the pupa after about 5–14 days, although this period depends on ambient temperature and also a stimulus. This is usually vibrations generated by a host's movements within its home, burrow or nest. If, however, animal shelters or houses are vacated adult fleas remain in their cocoons until their dwelling places are reoccupied. In some species, carbon dioxide emitted from hosts or a seasonal increase in humidity stimulates emergence. Adults may remain alive in their cocoons for *4–12 months*. This explains why people moving into buildings which have been vacated for many months may suddenly be attacked by large numbers of newly emerged very bloodthirsty fleas seeking their first blood-meal.

The life cycle from egg to adult emergence may be as short as 2–3 weeks for certain species under optimum conditions, but frequently the life cycle is considerably longer, taking many months.

Fleas avoid light and are usually found sheltering amongst the hairs or feathers of their hosts, or on people under their clothing or even in beds. During feeding fleas eject faeces composed of semi-digested blood of the previous meal and then excess blood ingested during the act of feeding. This mixture of partially digested and virtually undigested blood often marks clothing and bed linen of people heavily infested with fleas.

Although most species of fleas have favourite hosts they are not entirely host-specific; for example, cat and dog fleas (*Ctenocephalides felis* and *C. canis*) will readily feed on humans, as well as many wildlife species, especially in the absence of their normal hosts. Human fleas (*Pulex irritans*) feed on pigs, and rat fleas (*Xenopsylla* species) will attack people in the absence of rats. Most fleas will in fact bite other hosts in their immediate vicinity when their normal hosts are absent or scarce. Although feeding on less acceptable hosts keeps fleas alive, their fertility can be reduced. Fleas rapidly abandon dead hosts to find new ones, behaviour which is of profound epidemiological importance in plague transmission. While some fleas can *withstand* both considerable desiccation and prolonged periods of starvation, for example 6 months or more, when no suitable hosts are present, cat and dog fleas die within 10 days away from their hosts. On a host, fleas move by rapidly crawling, whereas off the host they jump more than crawl in their search for new hosts. Fleas can jump about 20 cm vertically and 30 cm or more horizontally. Such remarkable feats are achieved through a rubber-like protein called resilin. This is very elastic and can become highly compressed, then rapid expansion of the compressed state gives the power for jumping.

11.3   Medical importance

11.3.1 Flea nuisance

Although fleas can be important vectors of disease, the most widespread complaint about them concerns the annoyance of their bites, which may result in considerable discomfort and irritation. The most common nuisance flea is the cat flea, *Ctenocephalides felis* (four subspecies of *C. felis* are recognized, the true cat flea being *C. felis felis*, but for convenience it will be referred to as just *C. felis*). Of lesser importance as a pest is the dog flea, *Ctenocephalides canis*, and more rarely the human flea, *Pulex irritans*. The cat flea has become the most common flea on dogs.

Fleas frequently bite people on the ankles and legs, but at night a sleeping person may be bitten on any part of the body. People that become hypersensitive to flea bites can suffer from dermatitis, and inhalation of flea faeces can cause allergies. Children under 10 years tend to experience greater discomfort from flea bites than older people.

Because fleas are difficult to catch this increases the annoyance they cause. People attacked by fleas frequently spend sleepless nights alternately scratching themselves and trying to catch the fleas.

11.3.2 Plague

Bubonic plague, caused by *Yersinia pestis*, is a *zoonosis*, being primarily a disease of wild animals, especially rodents. About 203 rodent species and 14 lagomorphs (e.g. hares and rabbits) have been shown to harbour plague bacilli. The transmission cycle of plague between wild rodents, such as gerbils, marmots, voles, chipmunks and ground squirrels, is termed *sylvatic*, campestral, rural or enzootic plague. Many different species of fleas bite rodents and maintain plague transmission amongst them. When people such as fur trappers and hunters handle these wild animals there is the risk that they will get bitten by rodent fleas and become infected with plague.

An important form of plague is *urban plague*. This describes the situation when plague circulating among wild rodents has been transmitted to commensal rats (e.g. the black rat, *Rattus rattus* and brown (Norwegian) rat, *R. norvegicus*). It is maintained in the rat population by fleas such as *Xenopsylla cheopis* (Europe, Asia, Africa and the Americas), *X. astia* (South-east Asia) and *X. brasiliensis* (Africa, South America and India). When rats are living in close association with people, such as in rat-infested slums, fleas normally feeding on rats may bite humans. This commonly happens when rats are infected with plague and rapidly develop an acute and fatal septicaemia. On their death infected fleas leave the rats and feed on humans. In this way bubonic plague is spread by rat fleas to human populations.

The most important vector is *Xenopsylla cheopis* but other fleas, including *X. astia* and *X. brasiliensis*, more rarely *Nosopsyllus fasciatus* and even more rarely *Leptopsylla aethiopica*, are plague vectors in some areas. However, the last two rodent fleas are reluctant to feed on people and so rarely transmit the disease to humans. In addition to humans becoming infected by the bites of fleas that previously fed on infected rats, plague can also be spread from person to person by fleas, such as *Xenopsylla* species and *Pulex irritans*, feeding on a plague victim then on another person. This latter method, however, appears to play a minor role in transmission. *Pulex irritans* may play a more important role in transmission than previously considered, especially in areas not having *X. cheopis*, but it seems that transmission is mainly mechanical, that is through contaminated flea mouthparts.

It is important to understand the methods by which fleas transmit plague. Plague bacilli sucked up by male and female fleas during blood-feeding are passed to the stomach, where they multiply greatly and extend forwards to invade the *proventriculus*. In some species, especially those in the genus *Xenopsylla*, further multiplication in the proventriculus may result in it becoming partially, or more or less completely, *blocked*. In a partially blocked flea, because the proventriculus is not functioning normally, some of the blood that has been sucked into the stomach is regurgitated along with plague bacilli into the host (Fig. 11.5). With completely blocked fleas, blood is sucked up from a host with considerable difficulty about as far as the proventriculus, where it mixes with plague bacilli and is then regurgitated back into the host. Blocked fleas soon become starved and repeatedly bite in attempts to get a blood-meal, and therefore are epidemiologically potentially very dangerous.

Another, but less important, method of infection is by the flea's faeces being rubbed into abrasions in the skin or coming into contact with mucous membranes. Plague bacilli can remain infective in flea faeces for as long as three years. Occasionally the tonsils become infected with plague bacilli due to crushing infected fleas between the teeth.

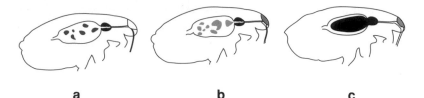

a   **b**   **c**

Figure 11.5 Development of *Yersinia pestis* in *Xenopsylla cheopis*: (a) early stage with proventriculus functioning normally; (b) partial blocking, with proventriculus failing to form an efficient valve mechanism; (c) completely blocked flea, with blood not being able to enter the stomach. (Courtesy of Miss M. A. Johnson, and Blackwell Publishing, Oxford, publishers of *Entomology for Students of Medicine* (1962) by R. M. Gordon and M. M. J. Lavoipierre.)

In septicaemic plague the bacilli are in the blood, and in pneumonic plague they invade the lungs. Very occasionally septicaemic plague is transmitted mechanically by fleas.

11.3.3 Murine typhus

Murine, flea-borne or endemic typhus is caused by *Rickettsia typhi*, the rickettsiae being ingested by the flea with its blood-meal. In the gut the rickettsiae multiply, but unlike plague bacilli they do not block the proventriculus. Transmission occurs when infected *faeces* are rubbed into abrasions or come into contact with delicate mucous membranes, and also by the release of rickettsiae from crushed fleas. Faeces remain *infective* for many months to a year or more; under laboratory conditions they have remained infective for 4.5–9 years! Murine typhus is essentially a disease of rodents, particularly rats such as *Rattus rattus* and *R. norvegicus*. It is spread among rats and other rodents by *Xenopsylla* species, especially *X. cheopis*, but also by *Nosopsyllus fasciatus* and *Leptopsylla segnis*. A few ectoparasites which are not fleas are vectors, such as the spined rat louse (*Polyplax spinulosa*) and possibly the cosmotropical rat mite (*Ornithonyssus bacoti*).

People become infected mainly by being bitten by *Xenopsylla cheopis*, but occasionally *Nosopsyllus fasciatus*, *Ctenocephalides canis*, *C. felis* and *Pulex irritans* may be involved. *Leptopsylla segnis* does not bite humans, but it is possible that murine typhus is sometimes spread to people by an aerosol of this flea's infective faeces.

Rickettsiae of murine typhus can pass to the flea's ovaries and subsequently to the eggs, larvae and adults, that is *transovarial* transmission. But whether this is epidemiologically important remains uncertain.

Rickettsia felis, very similar to *R. typhi*, has been isolated in the USA from opossums (*Didelphis* species) and cat fleas which feed on them. It seems there is an urban type of murine typhus involving rats and rat fleas (described above) and a rural type involving opossums and cat fleas (*C. felis*) which can also be transmitted to humans. The virus can be maintained in cat fleas by vertical transmission for many generations.

11.3.4 Cestodes

Dipylidium caninum is the commonest tapeworm of dogs and cats and occasionally occurs in children. It can be transmitted by fleas to both pets and humans as follows. Tapeworm proglottids containing eggs excreted by a pet crawl away from the host and dry on exposure to air. Larval fleas feeding on organic debries in host bedding bite into the dried proglottids, releasing the eggs, which they then swallow. Larval worms hatching from the ingested eggs penetrate the gut wall of the larval flea and enter the body cavity (coelom). They remain trapped here before passing to the pupa and finally to the adult flea, where they encapsulate and become cysticercoids (infective larvae). Animals become infected by licking their coats during grooming and swallowing infected adult fleas. Similarly, young children fondling and kissing dogs and cats can become infected by swallowing cat and dog fleas, or by being licked by dogs which have crushed infected fleas in their mouths, thus liberating the infective cysticercoids.

The rat tapeworms *Hymenolepis diminuta* and *H. nana* have similar life cycles.

11.3.5 Less important diseases

Cat-scratch disease (*Bartonella henselae*) is transmitted among cats by cat fleas. It seems that cats' claws are contaminated with *Bartonella*-infected flea faeces and that transmission to humans is mainly by cat scratches. Tularaemia (*Francisella tularensis*) may occasionally be transmitted to humans by flea bites. Ticks, however, are the main vectors, and it must be stressed that the role of fleas as vectors of these two infections is minimal.

11.4   *Tunga penetrans*

11.4.1 External morphology

Tunga penetrans is found in the West Indies, Central and South America and sub-Saharan Africa including Madagascar. It sometimes occurs in travellers returning home from these regions. *Tunga penetrans* is sometimes referred to as the chigoe, jigger flea or sand flea. *Tunga penetrans* does not transmit any disease but is a nuisance because females become imbedded in the skin.

Adults of both sexes are exceedingly small, only about 1 mm long (Fig. 11.6a). They have *neither* genal nor pronotal combs and are easily separated from other fleas of medical importance by their very *compressed*

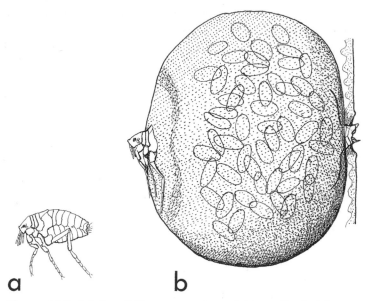

a   b

Figure 11.6 Adults of *Tunga penetrans* (chigoe flea): (a) non-gravid female; (b) gravid female with enormously swollen abdomen full of eggs which is embedded in the skin of a host. Note (on right) that the tip of the abdomen projects from the host's skin to the exterior.

first three (thoracic) segments, and by the paucity of spines and bristles on the body.

11.4.2 Life cycle and medical importance

Eggs of *Tunga penetrans* are dropped onto the floor of houses or on the ground outside, and hatch within about 3–4 days. Larvae inhabit dirty and dusty floors or dry sandy soils, especially in areas frequented by hosts of the adult fleas. There are two instars, not three as in most flea species, and under favourable conditions larval development is completed within about 10–14 days; the pupal period lasts about 5–14 days.

Newly emerged adults jump and crawl on the ground until they locate a host, usually a person or pig. ***Both*** sexes feed on blood, but whereas the male soon leaves the host after taking a blood-meal, the female, after being fertilized, 'burrows' into the skin where it is soft, such as between the toes or under toe-nails. Strictly speaking she does not burrow, but the host's skin envelopes her. Other areas of the foot, including the sole, may also be invaded. In people habitually sitting on the ground, such as beggars or infants, the buttocks may be infested, and particularly large infestations have been recorded from leprosy patients. In heavily infested individuals the arms, especially the elbows, may also be attacked, and occasionally the females 'burrow' into the soft skin around the genital region. The entire flea, with the exception of the tip of the abdomen bearing the anus, genital opening and large respiratory spiracles, becomes *completely buried* in the host's skin, and here she continues to feed. The area surrounding the embedded flea becomes very itchy and inflamed, and secondary infections may become established, resulting in ulcerations and accumulation of pus. While the blood-meal is being digested, the abdomen distends to an enormous size and after 8–10 days the flea is both the shape and size (6 mm) of a small pea (Fig. 11.6b). The abdomen now contains thousands of minute eggs, and over the next 7–14 days about 150–200 eggs are laid each day, most of which eventually fall to the ground and hatch after about 3–4 days. The life cycle from egg to adult usually takes 4–6 weeks but can be as little as 18 days.

After female fleas die they remain *embedded* in the host. This frequently causes inflammation and may result in secondary infections, which if ignored can lead to loss of the toes, tetanus, or even gangrene. Male fleas cause no such trouble as they do not 'burrow' into the skin.

Chigoe fleas are most common in people not wearing shoes, such as children. Up to 100 lesions have been found in the feet of a single child in endemic areas, and there are commonly 40 or more lesions per person. Because the fleas are feeble jumpers, wearing shoes is a simple, but in some communities relatively costly, method of reducing the likelihood of flea infestation.

Females embedded in the skin can be removed with fine needles under aseptic conditions, and the wounds should then be treated with antibiotics and dressed. They are best removed within the first few days of their becoming established, because when they have greatly distended abdomens, containing numerous eggs, they are difficult to extract without rupturing them, and this increases the risk of infections. Surgical extractions, however, are not easy and it may be better to treat lesions topically with lotions of 0.08% ivermectin or 0.2% metrifonate (trichlorfon) on two consecutive days to kill fleas *in situ*.

Pigs, in addition to humans, are commonly infested with *Tunga*, and may provide a local reservoir of infestation. Other animals such as cats, dogs and rats are also readily attacked.

11.5 Control of fleas

Repellents such as DEET or permethrin-impregnated clothing may afford some personal protection against fleas.

Insecticide resistance has been reported in cat fleas, human fleas and *Xenopsylla* species to one or more of the following categories of insecticides: organochlorines, organophosphates, carbamates, pyrethrins and pyrethroids. Nevertheless insecticides remain the main tool for flea control, although there is increasing reliance on insect growth regulators (IGRs).

11.5.1 Rodent fleas

Rodent fleas can be killed by applying insecticidal dusts such as ben-diocarb, propoxur, diazinon, malathion, pirimiphos-methyl, deltamethrin, permethrin or cypermethrin to rodent burrows and their nearby runways. Rodent fleas in houses can be controlled by spraying floors with bendio-carb, malathion, pirimiphos-methyl or pyrethroids, or with foggers containing permethrin or pirimiphos-methyl.

Controlling fleas in urban outbreaks of plague or murine typhus requires extensive and well-organized insecticidal operations. While insecticides are being applied rodenticides formulated as baits, such as the anticoagulants warfarin and brodifacoum, can be used to kill rodents. However, if fast-acting anticoagulants such as bromadiolone and chlorophacinone are used, then it is *essential* to apply these several days after insecticidal applications. Otherwise rodents will be killed before their fleas are killed, and the fleas will then bite other mammals including people, which may result in increasing disease transmission. Where there is resistance to warfarin and other anticoagulants calciferol, a fast-acting rodenticide, can often be substituted.

Although IGRs appear to have good potential for control of plague fleas they have received relatively little evaluation.

11.5.2 Cat and dog fleas

Cat and dog fleas (*Ctenocephalides felis* and *C. canis*) can be detected by examining the fur on the neck or stomach of the hosts. Powders, sprays or shampoos containing propoxur, malathion, chlorpyrifos, permethrin or deltamethrin can be applied to the animal's fur. Dusts are safer than liquids because they are less likely to be absorbed through the animal's skin and cause unpleasant side-reactions. Alternatively insect growth regulators (IGRs) such as pyriproxyfen and methoprene can be applied to the animal's skin.

A simple, but often not very effective, procedure is to use a plastic pet collar impregnated with propoxur or bendiocarb or an IGR. Flea collars remain effective for 2–3 months but sometimes up to a year for those containing methoprene, fenoxycarb or pyriproxyfen. Alternatively insecticides such as permethrin or imidacloprid (a nicotinoid insecticide) can be formulated as a 'spot-on' solution that is applied to the pet's skin in just one small area. Insecticides disperse over the entire body, suspended in skin oils, so fleas feeding anywhere on the host are killed. One treatment lasts for about 3–4 weeks. When IGRs such as cryomazine or lufenuron are administered orally once a month this gives prolonged ovicidal activity. Fleas feeding on the animals lay non-viable eggs for 2–3 weeks, and effective flea control may last for about 4–6 weeks.

However, an important consideration is that most fleas are found away from the host, not on it. Typically there may be only about 25 adult fleas on a cat, but on the floor and bedding apart from a few adult fleas there may be 500 cocoons and as many as 3000 larvae and 1000 eggs. Clearly, control measures should also be applied to beds, kennels, and other places where pets sleep. These items should either be treated with insecticidal powders or lightly sprayed with malathion, chlorpyrifos, one of the pyrethoids or an IGR. Duration of effective control depends on the types of materials sprayed (e.g. earthen or wooden floors, synthetic or woollen carpets), but IGRs generally remain effective longer than conventional insecticides, some giving good control for about 2–4 months. Because pupae are not very susceptible to insecticides, treatments may have to be repeated after two weeks, or when allowed by the label accompanying the insecticide, to kill newly emerged adult fleas.

Vacuum cleaning floors, carpets and pets' bedding can also be very effective in removing the immature stages of fleas.

Further reading

Azad, A. F. (1990) Epidemiology of murine typhus. *Annual Review of Entomology*, **35**, 553–69.

Azad, A. F. and Beard, C. B. (1998) Rickettsial pathogens and their arthropod vectors. *Emerging Infectious Diseases*, **4**, 179–86.

Eisele, M., Heukelbach, J., van Marck, E. *et al.* (2003) Investigations on the biology, epidemiology, pathology, and control of *Tunga penetrans* in Brazil: I. Natural history in man. *Parasitology Research*, **49**, 557–65.

Gage, K. L. and Kosoy, Y. (2005) Natural history of plague: perspectives from more than a century of research. *Annual Review of Entomology*, **50**, 505–28.

Gratz, N. G. (1999) Control of plague transmission. In *Plague Manual*: *Epidemiology, Distribution, Surveillance and Control*. Geneva: World Health Organization, pp. 97–134.

Hechemy, K. E. and Azad, A. F. (2001) Endemic typhus, and epidemic typhus. In *The Encyclopedia of Arthropod-Transmitted Infections of Man and Domesticated Animals*, ed. M. W. Service. Wallingford: CABI, pp. 165–9 and 170–4.

Heukelbach, J., Costa, A. M. L., Wilcke, T. and Mencke, N. (2004) The animal reservoir of *Tunga penetrans* in severely affected communities of north-east Brazil. *Medical and Veterinary Entomology*, **18**, 329–35.

Heukelbach, J., Franck, A. and Feldmeier, H. (2004) High attack rate of *Tunga penetrans* (Linnaeus 1758) infestation in an impoverished Brazilian community. *Transactions of the Royal Society of Tropical Medicine and Hygiene*, **98**, 43–4.

Hinkle, N. C., Koehler, P. G., Kern, W. H. and Patterson, R. S. (1991) Hematophagous strategies of the cat flea (Siphonaptera: Pulicidae). *Florida Entomologist*, **73**, 377–85.

Hinkle, N. C., Koehler, P. G. and Patterson, R. S. (1995) Residual effectiveness of insect growth regulators applied to carpet for control of cat flea (Siphonaptera: Pulicidae) larvae. *Journal of Economic Entomology*, **88**, 903–6.

Hinkle, N. C., Rust, M. K. and Reierson, D. A. (1997) Biorational approaches to flea (Siphonaptera: Pulicidae) suppression: present and future. *Journal of Agricultural Entomology*, **14**, 309–21.

Hirst, L. F. (1953) *The Conquest of Plague: a Study of the Evolution of Epidemiology*. Oxford: Clarendon Press.

Pugh, R. E. (1987) Effects on the development of *Dipylidium caninum* and on the host reaction to this parasite in the adult flea (*Ctenocephalides felis felis*). *Parasitological Research*, **73**, 171–7.

Rust, M. K. (2005) Advances in the control of *Ctenocephalides felis* (cat flea) on cats and dogs. *Trends in Parasitology*, **1**, 232–6.

Rust, M. K. and Dryden, M. W. (1997) The biology, ecology, and management of the cat flea. *Annual Review of Entomology*, **42**, 451–73.

Sachse, M. M., Guldbakke, K. K. and Khachemoune, A. (2007) *Tunga penetrans:* a stowaway from around the world. *Journal of the European Academy of Dermatology and Venereology*, **21**, 11–16.

Schriefer, M. E., Sacci, J. B., Taylor, J. P., Higgins, J. A. and Azad, A. F. (1994). Murine typhus: updated roles of multiple urban components and a second typhuslike rickettsia. *Journal of Medical Entomology*, **31**, 681–5.

Scott, S. and Duncan, C. J. (2001) *Biology of Plagues: Evidence from Historical Populations*. Cambridge: Cambridge University Press.

Traub, R. and Starcke, H. (eds.) (1980) *Fleas*. Proceedings of the International Conference on Fleas, Ashton Wold, Peterborough, UK, 21–25 June 1977. Rotterdam: Balkema.

12

Sucking lice (Anoplura)

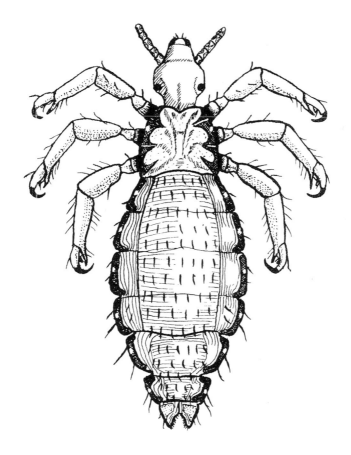

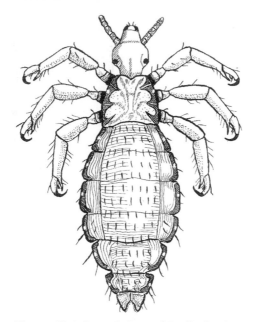

Figure 12.1 Dorsal view of *Pediculus humanus* (body louse). The head louse, *Pediculus capitis*, looks virtually identical.

Three types of blood-sucking lice occur on humans, the body louse (*Pediculus humanus*), the head louse (*Pediculus capitis*) and the pubic or crab louse (*Pthirus pubis*). Morphologically the body and head lice are virtually **indistinguishable**. In the laboratory the two can interbreed but very rarely do so in nature, and here they are treated as two distinct species, although many regard the head louse as a subspecies of the body louse. All three species of lice have a more or less worldwide distribution, but they are often more common in temperate areas.

Body lice are vectors of louse-borne typhus (*Rickettsia prowazekii*), trench fever (*Bartonella quintana*) and louse-borne relapsing fever (*Borrelia recurrentis*).

12.1　The body louse (*Pediculus humanus*)

12.1.1　External morphology

Adults are small, pale beige or greyish wingless insects, with a soft but rather leathery integument, and are **flattened** dorsoventrally (Fig. 12.1, Plate 17). Males measure about 2–3 mm and females about 3–4 mm. The head has a pair of small black eyes and a pair of short five-segmented antennae. The three thoracic segments are fused together and the legs are stout and well developed. The short thick tibia has apically a small **spine**

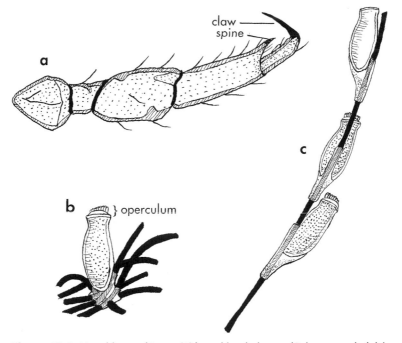

Figure 12.2 Head louse (*P. capitis*) and body louse (*P. humanus*): (a) leg of a body louse, showing tarsal claw and tibial spine; (b) unhatched egg of a body louse glued to fibres of clothing; (c) one hatched (upper) and two unhatched (lower) eggs of the head louse cemented to a hair. For convenience these three eggs are shown very close together, but in practice they are rarely this close on a head hair.

on its inner side, and the short tarsus ends in a curved *claw* (Fig. 12.2a). Hairs of the host, or clothing, are gripped between this tibial spine and tarsal claw.

Mouthparts of the louse differ from those of most blood-sucking insects in that they do not form a projecting proboscis, but consist of a sucking snout-like projection called the *haustellum*, which is armed on the inner surface with minute teeth that grip the host's skin during feeding. Needle-like stylets are thrust into the skin and saliva injected into the wound to prevent blood from clotting. Blood is sucked up and passes into the stomach for digestion.

The lateral margins of the abdominal segments are sclerotized and often appear darker than the rest of the segments.

In males there are dark transverse bands on the dorsal surface of the abdominal segments and the tip of the abdomen is rounded, whereas in females it is *bifurcated* and used to grip fibres of clothing during egg laying.

12.1.2 Life cycle

Both sexes take blood-meals, and feeding occurs at any time during the day or night. Both adult and immature stages live permanently on humans, clinging mainly to fibres of their clothing and usually only to body hairs during feeding. Female lice glue about 6–10 eggs per day very firmly on clothing fibres, especially on those along the seams of underclothes, such as vests and pants, but also on shirts and very occasionally on body hairs. The egg, commonly called a *nit* (though some apply this term only to the hatched egg), is oval, white, about 1 mm long, and has a distinct *operculum* (cap) containing numerous small perforations which give the egg the appearance of a minuscule pepper pot (Fig. 12.2b). Intake of air through these holes not only supplies the tissues of the developing embryo with oxygen but aids hatching in the following way. Just prior to hatching, the fully developed louse within its eggshell swallows air, which distends the body against the eggshell, thus building up a back pressure causing the head of the louse to be pushed up against the operculum and forcing it off. Female lice live for 2–4 weeks and may lay 150–300 eggs.

The egg stage lasts about 5–11 days except that eggs on discarded clothing may not hatch until 2–3 weeks; or in cool conditions not at all. Because eggs cannot survive longer than four weeks there is little danger of acquiring body lice from clothing not worn for over a month.

Lice have a *hemimetabolous* life cycle. The louse hatching from the egg is termed a *nymph* and resembles a small adult. It takes a blood-meal and passes through three nymphal instars, and after about 7–14 days becomes an adult male or female louse. Duration of the nymphal stages depends on whether clothing is worn continuously. If it is discarded at night and nymphs are subjected to lower temperatures, this may slow their development. A louse usually takes 3–5 blood-meals a day from its host. The life cycle from egg to adult is about 2–3 weeks.

The body louse is an ectoparasite of humans. Away from humans, unfed lice die within 2–4 days, but blood-fed individuals may survive for 5–10 days. On their hosts, blood-feeding adults probably live for about 30 days. Lice are very sensitive to changes in temperature. They quickly abandon a dead person to seek new hosts. They also leave a person with a high temperature, being unable to feed at temperatures above about 40 °C.

A very heavily infested person may have 400–500 lice on their clothing and body. In very exceptional conditions some 20 000 lice have been recorded from a single person! Usually, however, less than 100 lice are found on any one individual, and many have considerably fewer than this.

Body lice are spread by close contact and are especially prevalent under conditions of overcrowding and in situations where people rarely wash or change their clothes. They are therefore commonly found on people in primitive jails, refugee camps and in trenches during wars, and also after disasters such as floods or earthquakes when people are forced to live in very overcrowded, and usually insanitary, conditions. People living in mountainous areas in East Africa, Ethiopia, Sudan, Burundi, Nepal, India and Andean regions of South America, where cold weather necessitates wearing several layers of clothes which may be rarely changed or washed, often have lice. In more developed countries body lice are found mainly on homeless people, and infestations may reach a peak in cold weather when several layers of woollen underclothes are worn.

12.1.3 Medical importance
Pediculosis

Presence of body, head or pubic lice on a person is sometimes referred to as *pediculosis*. The skin of people who habitually harbour large numbers of body lice may become pigmented and tough, a condition known as vagabond's disease, hobo disease or sometimes as morbus errorum.

Because lice feed several times a day, saliva is repeatedly injected into individuals and its toxic effects may cause weariness, irritability or a pessimistic mood: the person feels lousy. Some people develop allergies such as dermatitis or severe itching, or have a type of asthmatic bronchitis. Secondary infections such as impetigo can also result from large numbers of biting lice.

Louse-borne epidemic typhus

Rickettsiae of louse-borne typhus, *Rickettsia prowazekii*, are ingested with blood-meals taken by both male and female lice, and also by their nymphs. They invade the epithelial cells lining the stomach of the louse and multiply enormously, causing the cells to become greatly distended. About four days after the blood-meal the *gut cells* rupture and release the rickettsiae back into the lumen of the insect's intestine. Due to these injuries the blood-meal may seep into the haemocoel of the louse, giving the body a reddish colour. Rickettsiae are passed out in the *faeces* of the louse, and people become infected when these are rubbed or scratched into abrasions, or come into contact with delicate mucous membranes such as the conjunctiva. Infection can also be caused by *inhalation* of the very fine powdered dry faeces. Also, if a louse is crushed, such as by persistent scratching, rickettsiae in the gut are released and may cause infection through abrasions etc. Rickettsiae can remain infective in dried louse *faeces* for about 70 days.

Humans, therefore, become infected with typhus either by the faeces of the louse or by crushing it, not by its bite. An unusual feature of louse-borne epidemic typhus is that it is a disease of the louse as well as of humans. Rupturing of the intestinal epithelial cells caused by multiplication of the rickettsiae frequently kills the louse after about 8–12 days. This may explain why people suffering from typhus are sometimes found with remarkably few, or no, lice on their bodies or clothing. An outbreak of epidemic typhus occurred in Russia in 1997, but it mainly occurs in cool mountainous regions of Africa, Asia, and Central and South America.

People are usually considered to be the reservoir hosts of typhus. Asymptomatic carriers remain infective to body lice for many years. *Recrudescences* as Brill–Zinsser disease, many years after the primary attack, may occur in a person and lead to the spread of epidemic typhus.

Louse-borne epidemic relapsing fever

Borrelia recurrentis is ingested with the louse's blood-meal from a person suffering from epidemic relapsing fever, but within about 24 hours all spirochaetes have disappeared from the lumen of the gut. Many have been destroyed, but the survivors have passed through the stomach wall to the *haemocoel*, where they multiply to reach enormous numbers after 10–12 days. The only known way someone can be infected is by the louse being crushed and the released spirochaetes entering the body through abrasions or mucous membranes. However, recently laboratory trials have shown that faeces of infected lice can contain live *B. recurrentis*. The habit of crushing lice between the finger-nails, or the less desirable habit of killing them by cracking them with the teeth, is clearly dangerous if lice are infected with relapsing fever or typhus. Louse-borne relapsing fever has disappeared from Europe but remains common in Central and East Africa, Sudan, Ethiopia, Afghanistan and Peru.

The method of transmission of epidemic relapsing fever must make it very rare for more than one person to be infected by any one infected louse. Hence epidemics of louse-borne relapsing fever will rarely occur unless there are large louse populations.

Trench fever

Trench fever is a relatively uncommon and non-fatal disease which was first recognized during World War I (1914–18) among soldiers in the trenches, and then reappeared in eastern Europe during World War II (1939–45). The disease disappeared again, only to reappear later in North America and Europe in the 1980s, occurring mainly in homeless people. It has also been reported in the 1990s and 2000s from Australia, Mexico, Peru, Japan, North Africa, Burundi and other sub-Saharan countries.

Trench fever is caused by *Bartonella quintana*. The bacteria are ingested by the louse during feeding and become attached to the walls of the gut cells, where they multiply; they do not penetrate the cells as do typhus rickettsiae and consequently they are not injurious to the louse. After 5–10 days the *faeces* are infected. Like typhus, the disease is conveyed to humans either by crushing the louse or by its faeces coming into contact with skin abrasions or mucous membranes. Bacteria persist for many months, possibly even a year, in dried louse faeces, and it is suspected that infection may commonly arise from *inhalation* of the dust-like faeces. The disease may be contracted by those who have no lice, but are handling louse-infected clothing contaminated with faeces.

12.1.4 Control

If louse-infested clothing is subjected to a minimum temperature of 70 °C for at least an hour body lice are killed. In epidemic situations, however, such measures may be impractical and immediate reinfestation may occur, so insecticides are usually used for louse control.

Lice are killed when insecticidal dusts, such as 5% carbaryl, 1% propoxur, 1% malathion or 0.5% permethrin, mixed with an inert carrier (e.g. talc), are blown by a plunger-type duster between the body and underclothes. However, checks should firstly be made to determine whether lice in the area have developed resistance to any of the insecticides to be used.

Impregnation of clothing with a pyrethroid insecticide may provide long-lasting protection against louse infestations, and such treated clothing may remain effective after several washings. Probably the best pyrethroid for this is 1% permethrin but 0.3–0.4% bioallethrin or 0.2–0.4% phenothrin can also be used.

Trials have shown that orally administered ivermectin kills body lice, and also head lice, but it is not yet universally approved for control of human lice.

12.2   The head louse (*Pediculus capitis*)

12.2.1 External morphology

Only very minor morphological differences separate body and head lice. In practice, these differences are not very important because lice found on clothing or on the body are invariably body lice, whereas those on the head are head lice.

12.2.2 Life cycle

The life cycle is very similar to that of the body louse except the eggs (*nits*) are not laid on clothes but are cemented to the hairs of the head, especially at their base (Fig. 12.2c), and normally hatch after 6–7 days. Usually a single egg is laid on each hair. The distance between the scalp and the furthest egg glued to a hair may provide an approximate estimate of the duration of infestation on the basis that a human hair grows about 1 cm per month. However, eggs may also be laid on long hairs when they are near or touching the scalp, so that unhatched eggs may be some distance from the base of the hair. Only very occasionally are eggs laid on hairs elsewhere on the body.

Most individuals have only 10–20 head lice, but in very severe infestations the hair may become matted with a mixture of nits, nymphs, adults and exudates from pustules resulting from bites of the lice. In such cases bacterial and fungal infections may become established and an unpleasant crust form on parts of the head, underneath which are numerous head lice. Empty, hatched *eggs* remain firmly cemented to the hairs of the head. A female lays about 6–8 eggs per day, amounting to about 50–150 eggs during her lifetime, which is about 2–4 weeks. Eggs hatch within 5–10 days and the duration of the nymphal stages is about 7–10 days. Away from people head lice die within 2–3 days.

As with body lice, dissemination of head lice is only by close contact, such as children playing together with their heads frequently touching, or when people are crowded together such as in prisons or refugee camps. Catching head lice from inanimate objects such as hats, scarves or chair backs is considered unlikely. Infestations are often, but not always, greater in women than men, and usually highest in children.

12.2.3 Medical importance
In many areas of the world head lice are a serious public health problem, and in many countries prevalence has been increasing. In some schools in the USA and the UK almost 50% of pupils have head lice. Often there are higher infestation rates in overcrowded homes and where hygiene is poor. There is little evidence that head lice are natural vectors of the diseases transmitted by body lice – for example, typhus epidemics are always associated with body lice – but they may occasionally be minor vectors in some outbreaks of louse-borne relapsing fever.

12.2.4 Control
Regular combing with an ordinary comb, although not removing the eggs, may reduce the number of nymphs and adults. A plastic *louse comb* with very closely set fine teeth is much better but may not remove all lice and their eggs. Alternatively, the head can be shaved.

Insecticidal formulations for louse control include dusts, emulsions and lotions. The choice depends on the availability of proprietary brands, preference of patients and costs. Although insecticidal dusts are efficient, they are not acceptable to most people because they give the head a greyish appearance signalling that the person has lice. Shampoos which are applied and then washed off after a few minutes are not usually very effective. Most commercial preparations for controlling head lice contain pyrethroids such as phenothrin, permethrin, bioallethrin or tetramethrin, or malathion or carbaryl, although in some countries carbaryl is no longer used against head lice because there is some evidence that it may pose a health risk. Some commercial preparations proclaim that lotions need only remain on the head for 10 minutes or 2 hours, but it is better to leave the insecticidal lotion on the head for about 12 hours, e.g. overnight, before washing it off.

Although some insecticides, such as malathion and permethrin, are reputed to be ovicidal, a *second* treatment after 7–10 days is recommended whatever insecticide is used, because it is difficult to kill all eggs with just a single application. None of the compounds will remove eggs cemented to the hairs, but these can be removed with a louse comb. As lice are readily transmitted between people it is recommended that all members of a household are treated, not just those detected as having lice.

Insecticide resistance, especially to the pyrethroids, is widespread and is the main reason for the increase in head lice infestations worldwide, but a novel head louse lotion (Hedrin) has recently become available. It is not a conventional insecticide but contains 4% dimeticone (silicone compound) which coats the lice and kills them. After 8 hours or more the hair can be washed. Retreatment after 7 days kills newly hatched lice from residual eggs. It is unlikely that resistance will evolve against dimeticone as it is not a typical insecticide.

Trials using orally administered ivermectin show the drug kills both head lice and body lice, but it is not yet universally approved for control of human lice.

12.3 The pubic louse (*Pthirus pubis*)

12.3.1 External morphology
The pubic louse is smaller (1.3–2 mm) than *Pediculus* species and is easily distinguished from them. In the pubic louse the body is nearly as broad as long, making it almost round. Whereas all three pairs of legs are more or less of equal size in the body and head louse, in the pubic louse the middle and hind-legs are much thicker than the front legs and have massive claws (Fig. 12.3). Presence of a **broad squat** body and *very large* claws, together with more sluggish movements, has resulted in the pubic louse being aptly called the crab louse.

12.3.2 Life cycle
Females lay about three eggs a day, totalling some 150–200 in their lifetime. They are slightly smaller than those of the body and head louse and are cemented to the coarse hairs of the genital and perianal regions of the body, and unlike head lice several eggs may be laid on a single hair. Pubic lice may be found on other areas of the body having coarse and not very dense covering of hair, for example, the beard, moustache, eyelashes, underneath the arms and occasionally on the chest. They are very rarely found on the head. Eggs take about 6–8 days to hatch and the duration of the three nymphal stages is about 10–17 days; consequently the life cycle is about 16–25 days.

Pubic lice are considerably less active than *Pediculus*. Infestation with crab lice is usually through sexual intercourse, and characteristically the French call them 'papillons d'amour'. However, it is wrong to suspect that this is the only method. Young children sleeping with parents can catch

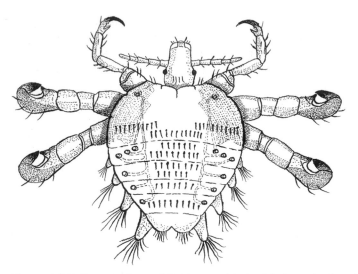

Figure 12.3 Dorsal view of *Pthirus pubis* (pubic louse), showing very large claws on mid- and hind-legs.

crab lice from them, and infestations can arise from discarded clothing, infested bedding, or even *rarely* from lavatory seats. Adults survive two days or less away from their hosts.

12.3.3 Medical importance
Although in the laboratory pubic lice can transmit louse-borne typhus, there is little evidence that under natural conditions they spread any disease, although it has been suggested that they have been responsible for typhus outbreaks in China. Severe allergic reactions (*pruritus*) can develop in response to their bites due to the injection of saliva and the deposition of faeces around the feeding sites. Small characteristic bluish spots (*maculae caeruleae*) may appear on infested parts of the body. Infestations of pubic lice are sometimes known as pediculosis pubis or phthiriasis (note that, unlike the generic name of the louse, this is spelt with a 'phth').

12.3.4 Control
Originally control involved shaving pubic hairs from the body, but this method has been replaced by the application of insecticidal lotions.

Basically insecticides used for head louse control can be used against pubic lice. Application of 1% permethrin or 5% malathion should kill nymphs and adults and possibly eggs, but a *second* application 7–10 days later is advisable in case some eggs are not killed. It may be advisable to treat all hairy areas of the body below the neck. *Aqueous*, not alcoholic, insecticidal solutions should be used, otherwise irritation may arise due to sensitivity of the genital region. Although resistance to pyrethroids has been reported, insecticide resistance in pubic lice appears to be rare. Infestations on the eyelashes can be treated by applying a small amount of a vaseline ointment containing pyrethrins daily over several days.

Only those medically trained should remove eggs from the lashes, which, however, will disappear as lashes fall out and are replaced by others.

Further reading

Burgess, I. F. (1998) Head lice: developing a practical approach. *The Practitioner*, **242**, 126–9.

Burgess, I. F. (2004) Lice and their control. *Annual Review of Entomology*, **49**, 457–81.

Burgess, I. F., Brown, C. M. and Lee, P. N. (2005) Treatment of head louse infestation with 4% dimeticone lotion: randomised controlled equivalence trial. *BMJ*, **330**, 1423–5.

Buxton, P. A. (1948) *The Louse: an Account of the Lice which Infest Man, Their Medical Importance and Control*, 2nd edn. London: Edward Arnold.

Chetwyn, K. N. (1996) An overview of mass disinfestation procedures as a means to prevent epidemic typhus. In *Proceedings of the 2nd International Conference on Insect Pests in the Urban Environment (ICIPUE)*, ed. K. B. Wildey. The Organising Committee of the ICIPUE, pp. 421–6.

Gratz, N. G. (1997) *Human Lice: Their Prevalence, Control and Resistance to Insecticides. A Review 1985–1997*. CTD/WHOPES/97.8. Geneva: World Health Organization.

Hill, N., Moor, G., Cameron, M. M. *et al.* (2005) Single blind, randomised, comparative study of the Bug Buster kit and over the counter pediculicide treatments against head lice in the United Kingdom. *BMJ*, **331**, 384–6.

Kristensen, M., Knorr, M., Rasmussen, A.-M., and Jespersen, J. B. (2006) Survey of permethrin and malathion resistance in human head lice populations from Denmark. *Journal of Medical Entomology*, **43**, 533–8.

Meinking, T., Burkhart, C. N. and Burkhart, C. G. (1999) Ectoparasitic diseases in dermatology: reassessment of scabies and pediculosis. *Advances in Dermatology*, **15**, 77–108.

Mumcuoglu, K. Y. (1996) Control of lice (Anoplura: Pediculidae) infestations: past and present. *American Entomologist*, **42**, 175–8.

Orkin, M. and Maibach, H. I. (eds.) (1985) *Cutaneous Infestations and Insect Bites*. New York, NY: Marcel Dekker. See Chapters 19–26 on lice.

Service, M. W. (ed.) (2001) *The Encyclopedia of Arthropod-transmitted Infections of Man and Domesticated Animals*. Wallingford: CABI. See K. E. Hechemy and B. E. Burton, *Bartonella quintana* and *Bartonella henselae*, pp. 70–3; K. E. Hechemy and A. F. Azad, Epidemic typhus, pp. 170–4; and D. A. Warrell, Louse-borne relapsing fever, pp. 295–9.

13

Bedbugs (Cimicidae)

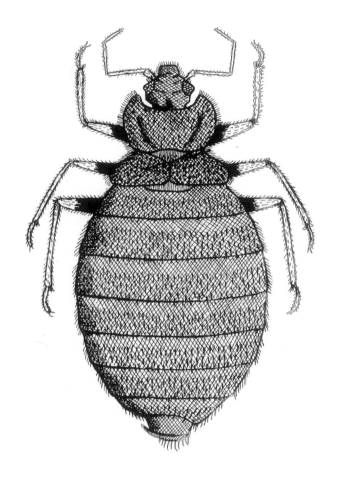

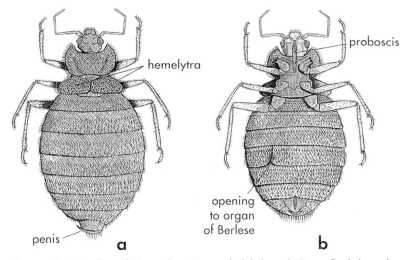

Figure 13.1 Bedbug (*Cimex hemipterus*): (a) dorsal view of adult male; (b) ventral view of adult female.

The family Cimicidae includes bedbugs, of which two common species feed on humans. *Cimex lectularius* is widely distributed in tropical and non-tropical countries while *C. hemipterus*, commonly called the tropical bedbug, is essentially a species of the Old and New World tropics although it can also occur in warm areas of some non-tropical countries, such as in Florida. It is not easy to separate these two species, but in *C. lectularius* the prothorax is generally 2.5 times as wide as long, whereas in *C. hemipterus* it is only about twice as wide as long. Also, the abdomen is more rounded in *C. lectularius* than in *C. hemipterus*.

A third species, *Leptocimex boueti*, found only in West Africa, bites bats and also people, but is much less important than the *Cimex* species.

Infestations of bedbugs have increased in many areas. They are not considered important vectors, but in addition to constituting a biting nuisance they have been reported as causing iron deficiency in infants.

13.1 External morphology

Adult bedbugs are oval, wingless insects which are *flattened* dorsoventrally (Fig. 13.1, Plate 18). They are about 5–7 mm long and when unfed pale yellow or brown, but after a full blood-meal they become a characteristically darker 'mahogany' brown. The head is short and broad and has a pair of prominent compound eyes, in front of which is a pair of four-segmented antennae. The proboscis is slender, and is normally held closely appressed under the head and prothorax, but when the bug takes a blood-meal it is swung forward and downwards (Fig. 13.2a). The *prothorax* is much larger than the meso- and metathorax and has distinct wing-like expansions. Two rudimentary and non-functional more or less oval wing pads, termed *hemelytra,* overlie the meso- and metathorax. The three pairs of legs are slender but well developed.

The abdomen has 11 segments but only 8 are readily visible. In adult males the tip of the abdomen is slightly more pointed than in females, while closer examination shows a small well-developed curved penis (Fig. 13.1a). In females there is a small incision ventrally on the left side of the apparent fourth abdominal segment (Fig. 13.1b). This opens into a special pouch (= sinus) called the mesospermalege or the organ of Berlese or organ of Ribaga, which collects and stores sperm. Because both male and female bugs bite it is not medically very important to distinguish the sexes.

13.2   Life cycle

Both sexes of bedbug take blood-meals and are equally important as pests. Feeding usually occurs at night on sleeping people, often just before dawn. If, however, bedbugs are starving they will feed during the day, especially in darkened rooms. Unlike lice, bedbugs do not remain on people but stay only to take blood-meals. Very occasionally, however, in temperate climates they remain on vagrants who rarely change their clothes. In the absence of people bedbugs will feed on other hosts, including rabbits, rats, mice, bats, poultry and other birds. During the day adults and nymphs are *inactive* and hide in dark and dry places, such as cracks and crevices in furniture, walls, ceilings or floorboards, underneath seams of wallpaper and between mattresses and beds. At night adults and nymphs emerge to feed on sleeping people, after which they return to their resting sites to digest their blood-meals. Bedbugs are gregarious and are frequently found in large numbers. They can move quite rapidly when disturbed.

Females lay about 2–3 eggs a day in cracks and crevices of buildings and furniture, but egg-laying ceases at 13 °C or lower. Eggs are about 1 mm long, pearly white or yellowish white, covered with a very fine and delicate mosaic pattern, and characteristically slightly *curved* anteriorly (Fig. 13.2b). Females live several weeks to many months, and occasionally a year or more, and during this time they may lay 150–500 eggs.

Eggs hatch after about 8–11 days, but within less than a week at higher temperatures. If, however, temperatures in houses are low, hatching may be delayed for several weeks, and such unhatched *eggs* can survive for three months. During hatching the small *operculum* (cap) is pushed from the anterior end of the egg, but often remains partially attached. Empty eggshells usually remain cemented in place after hatching. Newly hatched bedbugs (*nymphs*) are very pale yellow and resemble adults, but are much smaller (Fig. 13.2c). The life cycle is *hemimetabolous* and there are five nymphal instars, each of which takes one or more blood-meals. The nymphal period lasts 2–7 weeks, but is greatly extended in cool conditions

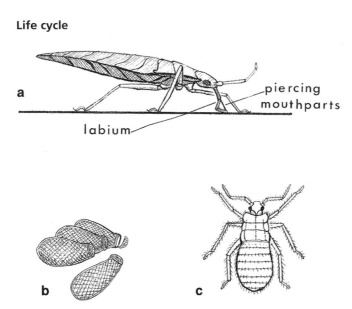

Figure 13.2 Bedbugs: (a) diagram of an adult with proboscis swung forward for blood-feeding; (b) one hatched and three unhatched eggs; (c) first-instar nymph.

or if regular blood-feeding is prevented by lack of hosts. The life cycle, from egg to adult (Fig. 13.3), can be just 3–4 weeks if temperatures are high and food plentiful, but is more usually 6–10 weeks. In the laboratory *adults* can live for four years, and survive more than a year without blood-feeding, but survival is dependent on temperature and humidity.

The method of mating in bedbugs is unique among insects. The penis penetrates the integument and enters the mesospermalege (organ of Ribaga or organ of Berlese) situated on the ventral surface of the female abdomen. Sperm introduced into this 'copulatory pouch' (= sinus) pass into the haemocoel and then ascend the oviducts to fertlize the eggs.

Bedbug *infestations* can be detected by the presence of live bugs, cast-off nymphal skins, and hatched and unhatched eggs, all of which may be found in cracks and crevices. In addition, small dark brown or black marks may be visible on bed sheets, walls and wallpaper: these are the bedbug's excreta and consist mainly of excess blood ingested during feeding. Houses with large bedbug infestations may have a characteristic rather sickly smell, but in practice this may not be apparent because the weak odour can be masked by stronger insanitary smells.

Because bedbugs lack wings they do not disperse far, although occasionally they crawl from one building to another. Bedbugs are usually spread to new houses by being introduced with furniture and bedding, or more rarely with clothing and hand baggage. Buying secondhand furniture can result in the introduction of bedbugs into houses.

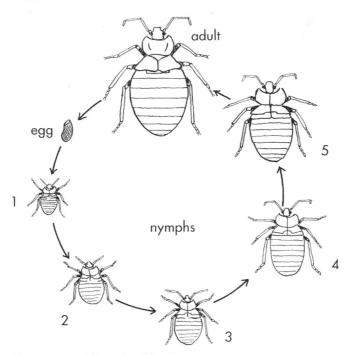

Figure 13.3 Life cycle of bedbugs.

13.3   Medical importance

Although hepatitis B virus and 27 other other pathogens have been recorded in bedbugs there is no evidence that they can transmit any infections to people. They are therefore not considered as vectors.

In areas with dilapidated buildings and poor hygiene standards, bedbug infestations can cause considerable distress. Some people show little or no reaction to their bites whereas others may suffer severe reactions and have sleepless nights. Repeated feedings of large numbers of bedbugs can cause iron deficiency in infants and some elderly people.

13.4   Control

Insect repellents and pyrethroid-impregnated bed-nets may give considerable personal protection against bedbugs.

Floors and walls of infested houses together with as much furniture as possible can be sprayed with 0.2–0.3% bendiocarb, 2% malathion, 1% pirimiphos-methyl or a range of pyrethroids such as 0.1% cypermethrin, 0.03% deltamethrin, 0.1% permethrin or 0.1–0.2% tetramethrin. If non-pyrethroids are used the addition of 0.1–0.2% pyrethrins or synthetic pyrethroids can help flush out bedbugs from their hiding places, so increasing their contact with the insecticide. Mattresses and wooden slats across beds can be dusted with insecticidal powders or lightly sprayed with insecticides, but must be aired afterwards to allow them to dry out completely before being re-used. Bedclothes and infants' mattresses should *not* be treated with insecticides.

Commercially available small insecticidal smoke generators containing permethrin or pirimiphos-methyl, which burn for up to 15 minutes, can be used to fumigate infested premises.

Insect growth regulators (IGRs) such as hydroprene have sometimes been used to control bedbugs, especially in the USA.

Further reading

Anon. (1973) *The Bedbug*, 8th edn. Economic Series 5. London: British Museum (Natural History).

Doggett, S. L., Geary, M. J., Lamond, P. and Russell, R. C. (2004). Bed bug management: a case study. *Professional Pest Manager*, **8** (6), 21–3.

Gooch, H. (2005) Hidden profits: there's money to be made from bed bugs – if you know where to look. *Pest Control*, **73** (3), 26–32.

Johnson, C. G. (1941) The ecology of the bedbug, *Cimex lectularius* L., in Britain. *Journal of Hygiene, Cambridge*, **41**, 345–461.

King, F. (1990) Mind the bugs don't bite. *New Scientist*, 27 January, 51–4.

Mayans, M. V., Hall, A. J., Inskip, H. M. *et al.* (1994) Do bedbugs transmit hepatitis B? *Lancet*, **343**, 761–3.

Reinhardt, K. and Siva-Jothy, M. T. (2007) Biology of bed bugs (Cimicidae). *Annual Review of Entomology*, **52**, 351–74.

Ryckman, R. E., Bentley, D. G. and Archbold, E. F. (1981) The Cimicidae of the Americas and Oceanic Islands: a checklist and bibliography. *Bulletin of the Society of Vector Ecologists*, **6**, 93–142.

Usinger, R. L. (1966) *Monograph of Cimicidae (Hemiptera–Heteroptera)*. Thomas Say Foundation, Vol. 7. Maryland: Entomological Society of America.

Venkatachalam, P. S. and Belavady, B. (1962) Loss of haemoglobin iron due to excessive biting by bed bugs: a possible aetiological factor in the iron deficiency anaemia of infants and children. *Transactions of the Royal Society of Tropical Medicine and Hygiene*, **56**, 218–21.

Weidhaas, D. E. and Keiding, J. (1982) Bed bugs. Mimeographed document WHO/VBC/82.857. Geneva: World Health Organization.

14

Triatomine bugs (Triatominae)

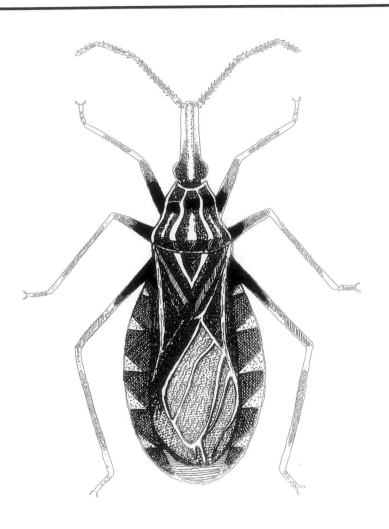

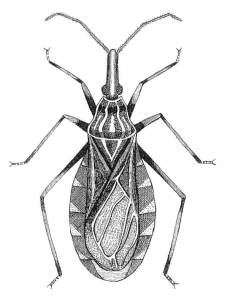

Figure 14.1 Adult *Rhodnius* species, an example of a triatomine bug.

Blood-sucking bugs in the family Reduviidae belong to the subfamily Triatominae and comprise more than 130 species in 15–17 genera (the number depending on whose authority is accepted). The most important are *Triatoma infestans*, *T. dimidiata*, *Rhodnius prolixus* and *Panstrongylus megistus*, all of which spread Chagas disease (*Trypanosoma cruzi*) in Central and South America. Some can also transmit *Trypanosoma rangeli*, a non-pathogenic organism.

Most Triatominae occur in the Americas, from the Great Lakes of the USA to southern Argentina, but 13 species are found in the Old World tropics. All medically important species are, however, confined to the southern USA, Central and South America. Triatomines are commonly called kissing-bugs, cone-nose bugs, vinchucas or barbeiros.

14.1 External morphology

Triatominae vary from 5 to 45 mm in length, but most are 20–30 mm long. They are easily recognized by their long *snout-like* head with a pair of prominent dark-coloured eyes, in front of which is a pair of *laterally* situated, long and thin four-segmented *antennae* (Fig. 14.1, Plate 19). The proboscis, sometimes called the rostrum, is relatively thin and straight and, as in bedbugs, lies closely appressed to the ventral surface of the head (Fig. 14.2a). However, when the Triatominae take a blood-meal the proboscis is swung forward and downwards (Fig. 14.2b).

The dorsal part of the first thoracic segment of the Triatominae consists of a very conspicuous triangular *pronotum*. The meso- and metathorax are

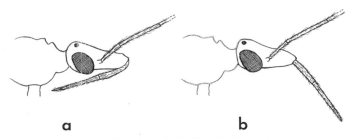

| a | b |

Figure 14.2 Lateral views of the head of a *Triatoma* species: (a) proboscis closely appressed to ventral side of the head; (b) proboscis swung forward in a blood-feeding position.

basal part   distal part

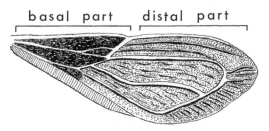

Figure 14.3 Fore-wing (hemelytron) of a triatomine bug, showing thickened basal part and more membranous distal part.

hidden dorsally by the folded fore-wings, called **hemelytra**. The basal part of each hemelytron is thickened and relatively hard, whereas the more distal part is membranous (Fig. 14.3). Hind-wings are entirely membranous, but when the bug is not flying they remain hidden underneath the hemelytra. The relatively long and slender legs end in paired small claws.

The 11-segmented abdomen is more or less oval in shape but is mostly hidden by the wings, except for the lateral margins, which are bent upwards slightly and are visible dorsally.

Triatominae are frequently a brown-black colour, but some species are more colourful, having contrasting yellow, orange, pink or red markings, usually as bands on the pronotum, basal part of the fore-wings, legs or abdominal margins.

14.2 Life cycle

Eggs are about 1.5–2.5 mm long, oval in shape, but have a slight constriction before the *operculum* (cap) (Fig. 14.4). They have a smooth or ornate shell which is pearly white, pink or yellowish depending on the species. Eggs are deposited in cracks and crevices in walls, floors, ceilings and furniture of houses, especially dilapidated mud-walled and thatched-roofed houses in rural areas, or slums at the edge of towns. Some species lay their eggs in rodent burrows and other shelters used by mammalian hosts upon which they feed. Avian feeders deposit their eggs in birds' nests or on leaves of

Figure 14.4 Two hatched and two unhatched triatomine eggs.

trees. Typically females lay 1–2 eggs a day. The total number laid varies from 50 to 1000, depending on the species, their longevity and the number of blood-meals they take, but it is usually 200–300.

The life cycle is *hemimetabolous*. Small pale *nymphs*, which resemble adults but lack wings, may hatch from eggs after only 10–15 days, but the incubation period may extend to 30 or 60 days. Newly emerged nymphs usually remain hidden in cracks and crevices for 2–3 days before they blood-feed. There are five nymphal instars, each requiring at least one blood-meal before it changes into the succeeding instar. Rudimentary wing pads are visible in the fourth and fifth nymphal stages, but only adults have fully developed wings. *Young nymphs* can ingest 6–12 times their own weight of blood, and their abdomens may become so greatly distended that they resemble blood-red balloons. Successive instars take relatively less blood, so that the fifth and last nymphal stage takes about 3–5 times its own weight of blood, while adults ingest 2–4 times their weight of blood. Adults of some species ingest 300–400 mg of blood at each meal and feed every 4–9 days! Sometimes hungry nymphs and adult bugs pierce the swollen abdomens of freshly engorged nymphs and take a blood-meal from them, without apparently causing harm.

Nymphs and adults of *both* sexes feed at night on their hosts, and *feeding* often lasts 10–25 minutes. People covered with blankets are bitten on any exposed parts of the body, but especially on the nose and around the eyes and mouth. Biting is usually relatively painless and does not awaken people, although some species cause considerable discomfort and there may be prolonged after-effects. Many bugs defecate during or soon after feeding, and this behaviour is very important in the transmission of Chagas disease. Presence of bugs in houses is often characterized by finding shed skins (exuvia) from moulting nymphs and streaks of whitish or dark faecal deposits on walls and furniture.

Because of the relatively long time required to digest their large blood-meals, the *life cycle* from egg to adult can take 3–10 months. With large triatomine species the life cycle may last 1–2 years. In the absence of hosts, older nymphs and adults can survive 4–6 months of starvation.

Triatomine bugs inhabit both forests and drier areas of the Americas. Many species feed on wild animals, such as armadillos, opossums, rats, mice, marsupials, ground squirrels, skunks, iguanas, bats, and also birds. Adults and nymphs are usually found in the burrows or nests of these animals. In addition to these sylvatic species, some bugs have become highly domesticated. They feed on donkeys, cattle, goats, horses, pigs, cats, dogs and especially chickens, which in some areas may be particularly important hosts, and of course humans. These domestic species often live in man-made shelters including houses, especially primitive ones made of wood, mud and thatch. Sylvatic species sometimes move into houses as forests are cut down and people occupy previously uninhabited areas. Some species are partly sylvatic and partly domestic in their feeding and resting habits.

If hosts vacate their shelters or homes, hungry nymphs crawl out and seek new hosts, whereas adults fly out to find new hosts and shelters. Some species are attracted by lights into houses.

14.3   Medical importance

14.3.1 Chagas disease

In rural areas of the Americas there may be hundreds of triatomine bugs in a house and this can be very stressful to the occupants, who will receive many bites during the night. Typically blood loss can exceed 2 ml per person per night, so it is not surprising that large bug populations can contribute to *anaemia*.

However, the main importance of the Triatominae is that they transmit *Trypanosoma cruzi*, the causative agent of Chagas disease, sometimes referred to as South American trypanosomiasis. *Trypomastigotes* ingested with a blood-meal undergo their entire development within the *gut* of the bug. They develop into *epimastigotes* and greatly multiply, and after 8–17 days become infective metacyclic trypomastigotes in the lumen of the hind-gut. *Blood-feeding* commonly lasts 10–25 minutes or longer, and during this time, or soon afterwards, many species of bugs excrete liquid or semiliquid faeces which may be contaminated with the metacyclic forms of *T. cruzi* derived from a previous blood-meal. People become infected when *excreta* is scratched either into skin abrasions or in the site of the bug's bite, or when it gets rubbed into the eyes or other mucous membranes. If the bug's bite produces local irritation causing the person to scratch, this facilitates infection. *Transmission* is not by the bite of the insect, only through its faeces.

About 70 triatomine species have been recorded naturally infected with *T. cruzi* but in practice only about a dozen species living in very close association with people, and therefore regularly feeding on them, are important vectors. Of these the principal ones are *Triatoma infestans* (southern South America), *Panstrongylus megistus* (mainly south-east Brazil), *Rhodnius prolixus* (mainly Honduras, Nicaragua, Colombia and Venezuela) and

T. dimidiata (Mexico to Colombia, Ecuador and Peru). Efficiency of a vector will depend on its speed of feeding and whether it defecates on a person during feeding.

Chagas disease is a **zoonosis**. *Trypanosoma cruzi* is essentially a parasite of wild animals, such as opossums (*Didelphis* species), armadillos (*Dasypus* species), many species of wild and urban rats and mice, squirrels, carnivores, monkeys and possibly bats, all of which may serve as **reservoir hosts**. The **bug** itself can also be a reservoir of infection, but in some areas humans are considered to be the principal reservoir host. Apart from acquiring *T. cruzi* through a bug's faeces, some animals become infected by eating the bugs or infected animals, such as carnivores, which have fed on rodents infected with *T. cruzi*. Rarely people can also aquire infection by eating infected meat (e.g. inadequately cooked opossums) or food contaminated with excrement of infected bugs.

Infection rates of Triatominae are often exceptionally high. For example, it is not uncommon to find infection rates of about 25% or even 50% or more. Even higher infection rates (78%) have been found in *Triatoma protracta* in California, but because this species very rarely bites people it is not considered a vector to humans. Vectors account for more than 80% of transmission; blood transfusion (17%) and congenital transmission (2%) also occurs.

14.3.2 *Trypanosoma rangeli*

Another trypanosome, *Trypanosoma rangeli*, occurring from Mexico to Brazil and apparently non-pathogenic in humans, is also transmitted by triatomines, especially by *Rhodnius prolixus*. In the vector the trypanosomes undergo dual development: some of the metacyclic infective forms migrate to the hind-gut while others penetrate the gut wall, pass across into the haemocoel and then migrate to the salivary glands. Humans are mainly *infected* by the bug's bite and only rarely by its faeces.

14.4   Control

Control of Chagas disease is mainly by spraying the interior surfaces of walls and roofs/ceilings of houses, outhouses, chicken sheds and goat pens with residual insecticides. Although fenitrothion (100 mg/m^2) is sometimes used pyrethroids are most commonly sprayed, particularly deltamethrin (25 mg/m^2), cyfluthrin (50 mg/m^2) and lambda-cyhalothrin (30 mg/m^2), and to a lesser extent cypermethrin (100 mg/m^2). Careful surveillance detects whether after a year there is reinfestation or foci of bugs that need further spraying. Resisistance to several insecticides including some pyrethrioids has been recorded, but presently this does not hinder control operations. Insecticidal smoke bombs which when lit dispense pyrethroid insecticides are sometimes used in houses to alleviate biting.

Bug populations can be reduced by making houses unattractive as resting sites: for example, by plastering walls to cover up cracks in which the bugs might hide, and by replacing dilapidated mud and thatched houses with those built of bricks or cement blocks and having corrugated metal roofs. However, because of the high costs of building new houses, rehousing has yet to be carried out on a large scale.

The above methods will destroy bugs resting in houses but are less effective against those resting in natural outdoor shelters. Such peridomestic populations can sometimes invade houses.

14.4.1 'Southern Cone Initiative'

In 1991 Argentina, Bolivia, Brazil, Chile, Paraguay and Uruguay, joined in 1996 by Peru, proposed a plan of action to eliminate the main vector, *Triatoma infestans*, chiefly by vector control methods such as insecticidal spraying of houses.

In 1997 Uruguay was declared free of Chagas disease transmission, as was Chile in 1999 and Brazil in 2005. Other countries, especially Bolivia and Peru, are moving towards interrupted transmission of Chagas. In 1997 two new initiatives were launched to reduce transmission, one in the Central American countries of Belize, Costa Rica, El Salvador, Guatemala, Honduras, Nicaragua and Panama, the other to cover the Andean countries of Colombia, Venezuela, Peru and Ecuador. However, as the main vectors in these two regions, *Triatoma dimidiata* and *Rhodnius prolixus*, are not so endophilic, indoor residual spraying is not so effective. At present there is little information on the success of these two initiatives.

The World Health Organization's goal is the elimination of Chagas disease in Latin America by 2010.

Further reading

Barrett, T. V. (1991) Advances in triatomine bug ecology in relation to Chagas' disease. *Advances in Disease Vector Research*, **8**, 143–76.

Beard, C. R., Cordon-Rosales, C. and Durvasula, R. V. (2002) Bacterial symbionts and their potential use in control of Chagas disease transmission. *Annual Review of Entomology*, **47**, 123–41.

Brenner, R. R. and Stoka, A. de la M. (eds.) (1988) *Chagas' Disease Vectors. I: Taxonomic, Ecological and Epidemiological Aspects*. Boca Raton, FL: CRC Press.

Brenner, R. R. and Stoka, A. de la M. (eds.) (1988) *Chagas' Disease Vectors. II: Anatomic and Physiological Aspects*. Boca Raton, FL: CRC Press.

Brenner, R. R. and Stoka, A. de la M. (eds.) (1988) *Chagas Disease Vectors. III: Biochemical Aspects and Control*. Boca Raton, FL: CRC Press.

Bryan, R. T., Balderrama, F., Tonn, R. J. and Dias, J. C. P. (1994) Community participation in vector control: lessons from Chagas' disease. *American Journal of Tropical Medicine and Hygiene*, **50** (suppl.), 61–71.

Carcavallo, R. U., Galíndez-Girón, I. G., Jurberg, J. and Lent, H. (eds.) (1999) *Atlas of Chagas' Disease Vectors in the Americas. Volume 3.* Rio de Janeiro: Fundación Oswaldo Cruz.

Feliciangeli, M. D., Campbell-Lendrum, D., Martinez, D., Gonzalez, D., Coleman, P. and Davies, C. (2002) Chagas disease control in Venezuela: lessons for the Andean region and beyond. *Trends in Parasitology,* **19,** 44–9.

Forattini, O. P. (1989) Chagas' disease and human behavior. In *Demography and Vector-Borne Diseases,* ed. M. W. Service. Boca Raton, FL: CRC Press, pp. 107–20.

Guhl, F. and Schofield, C. J. (eds.) (2004) *Proceedings of the International Workshop on Chagas Disease Surveillance in the Amazon Region.* Bogota: Universidad de los Andes.

Kingman, S. (1991) South America declares war on Chagas' disease. *New Scientist,* 19 October, 16–17.

Lent, H. and Wygodzinsky, P. (1979) Revision of the Triatominae (Hemiptera, Reduviidae), and their significance as vectors of Chagas' disease. *Bulletin of the American Museum of Natural History,* **163,** 123–520.

Moncayo, A. and Ortiz Yanine, M. I. (2006) An update on Chagas disease (human trypanosomiasis). *Annals of Tropical Medicine and Parasitology,* **100,** 663–77.

Schofield, C. J. (1988) Biosystematics of the Triatominae. In *Biosystematics of Haematophagous Insects,* ed. M. W. Service. Oxford: Clarendon Press, pp. 280–312.

Schofield, C. J. (1994) *Triatominae: Biology and Control.* Bognor Regis: Eurocommunica.

Schofield, C. J. and Dujardin, J. P. (1997) Chagas disease vector control in Central America. *Parasitology Today,* **13,** 141–4.

World Health Organization (2002) Control of Chagas disease. Second report of the WHO expert committee. *World Health Organization Technical Report Series,* **905,** 1–109.

Yamagata, Y. and Nakagawa, J. (2006) Control of Chagas disease. *Advances in Parasitology,* **61,** 129–65.

15

Cockroaches (Blattaria)

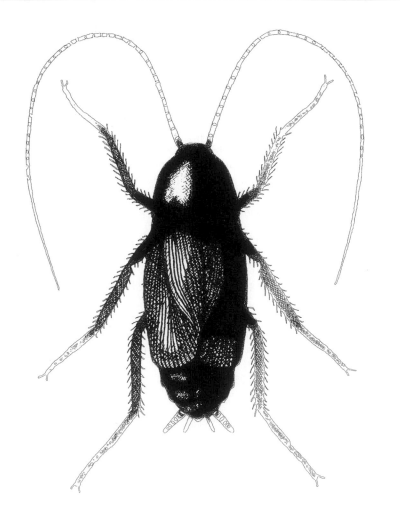

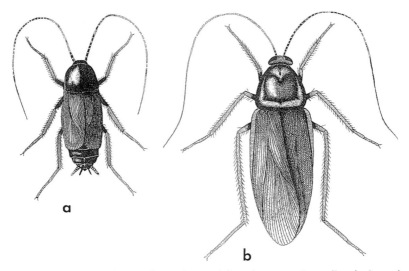

Figure 15.1 Adult cockroaches: (a) *Blatta orientalis* (oriental cockroach); (b) *Periplaneta americana* (American cockroach).

Cockroaches belong to the order Blattaria, and there are about 4000 species of which about 18 have become serious domestic pests. The most important medically are *Blattella germanica* (the German cockroach), *Blatta orientalis* (the oriental cockroach), *Periplaneta americana* (the American cockroach) and *Supella longipalpa* (the brown-banded cockroach). Cockroaches are sometimes called roaches or steambugs. They have an almost worldwide distribution.

Cockroaches aid in the mechanical transmission of various pathogenic viruses, bacteria and protozoans.

15.1 External morphology

The more common household pest species are chestnut brown or black, 10–50 mm long, flattened dorsoventrally, and have a smooth, shiny and tough integument. The head is small and sometimes almost hidden by the large, rounded *pronotum*. A pair of long and prominent filiform *antennae* arise from the front of the head between the eyes (Fig. 15.1, Plate 20). Cockroach mouthparts are developed for chewing, gnawing and scraping; they cannot suck blood. Both sexes have two pairs of wings. In some household species *wings* in the female are shorter than those of the male, and female *Blatta orientalis* has very small non-functional wings. The cockroach fore-wings, called *tegmina*, are thick and leathery. They are not used in flight but are protective covers for the membranous hind-wings, which when not in use are folded shut, fan-like, over the body. Although the hind-wings are used for flying cockroaches rarely fly. The well-developed legs are covered with prominent small spines and bristles; the five-segmented tarsi end in a pair of claws.

The segmented abdomen is more or less oval, and depending on species is either completely or partly hidden by the folded overlapping wings. In both sexes a pair of prominent segmented pilose *cerci* arise from the last abdominal segment (Fig. 15.2a), although in some species they are hidden from view by the wings.

Cockroaches are distinguished from beetles (order Coleoptera) by having the *fore-wings* placed over the abdomen in a closed scissor-like manner. In beetles the fore-wings (elytra) meet dorsally to form a distinct line down the centre of the abdomen. In addition, the elytra of beetles are generally thicker than the tegmina of cockroaches.

15.2   Life cycle

Cockroaches like warmth and in temperate countries hide behind radiators and hot-water pipes during the *day*. In warm countries they are found in almost any dark place, such as cesspits, septic tanks, sewers, refuse tips, dustbins, cupboards, underneath chairs, tables, sinks, baths and beds, behind refrigerators and cooking stoves. They are common in kitchens, especially when food is left out overnight, and in restaurants, hotels, bakeries, breweries, laundries and aboard ships. They are also frequently common in hospitals. Cockroaches are *nocturnal* and are rarely seen during the day unless they are disturbed from their hiding places. They become very active at night, running over floors, tables and other furniture to seek food. When lights are suddenly switched on cockroaches can be both seen and heard scuttling along.

They are *omnivorous* and voracious feeders, eating any type of food. They also eat paper, clothes, particularly starched ones, books, wallpaper, dried blood, sputum, excreta, dead insects, and almost any animal or vegetable matter. They may gnaw the finger-nails and toe-nails of babies and sleeping or comatose people, and even infest the hair of vagrants. Cockroaches habitually disgorge partially digested meals and deposit their excreta on almost anything, including food. They can *live* for 5–10 weeks without water and for many months without food, but this is not an important limiting factor because they very rarely occur in areas where food is not available. Young nymphs, however, may die within about 7–10 days in the absence of food.

Eggs are laid encased in a brown bean-shaped case called an *ootheca* (Fig. 15.2b), which can contain 12–50 eggs but typically 18–40. Cockroaches, especially *Blattella germanica*, are often seen running around with an ootheca partly protruding from the tip of the abdomen (Fig. 15.2a). Oothecae are deposited in cracks and crevices in dark and secluded places, and in some species are cemented to surfaces, such as the undersides of tables, chairs and beds.

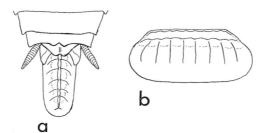

Figure 15.2 (a) Ootheca protruding from the abdomen of *Blatella germanica* (German cockroach); (b) lateral view of a typical cockroach ootheca.

Cockroaches live for many months to two or more years, and during this time females will lay, varying according to species, 4–90 oothecae.

Cockroaches have a *hemimetabolous* life cycle. Nymphs hatch from eggs after about 1–3 months, the time depending on both temperature and species. Young nymphs are very pale and delicate versions of the adults, whereas the older ones are progressively darker and more like the adults. Nymphs are wingless; the wings gradually develop with ensuing nymphal instars, but only adults may have fully developed wings. The number of nymphal stages is variable within and between species. There are commonly between *five and seven nymphal stages* but there may be as many as 13, such as sometimes in *Periplaneta americana*. Duration of the nymphal stage varies according to temperature, abundance of food and species. It commonly lasts 2–3 months in *Blattella germanica*, 12–15 in *Blatta orientalis*, and 10–14 in *Periplaneta americana*, but occasionally up to 23 months.

Cockroaches spread very rapidly from infested houses to adjoining ones. They often gain entry by climbing up water pipes and waste pipes. They are also spread as oothecae, nymphs and adults in furniture and other belongings.

15.3 Medical importance

15.3.1 Allergies
Only relatively recently has the importance of cockroach allergies been recognized. About half of asthmatics are allergic to cockroaches, their cast-off skins or excreta, while about 10% of non-asthmatic people will exhibit cockroach allergies. Symptoms include sneezing, skin reactions, sore eyes and in extreme cases shortness of breath. The allergic rate is second only to that caused by house-dust mites (Chapter 20).

15.3.2 Infectious agents
Because of their dirty habits of feeding indiscriminately on both excreta and foods, and of excreting and regurgitating partially digested meals over

food, the presence of cockroaches in houses, hotels and hospitals is highly undesirable.

Most parasitic infections isolated from cockroaches are also spread directly from person to person without the aid of intermediary insects, so it is usually difficult to prove that cockroaches are responsible for any disease outbreak. Nevertheless because of their insanitary habits they have been suspected as aiding the transmission of various pathogens. For example, they are known to carry poliomyelitis virus, *Entamoeba histolytica*, *Toxoplasma gondii*, *Escherichia coli*, *Staphylococcus aureus*, *Klebsiella pneumoniae*, *Shigella dysentariae* and *Salmonella* species, including *S. typhi* and *S. typhimurium*. Eggs of the nematode *Enterobius vermicularis*, which is an extremely common worm in humans, can also be carried by cockroaches.

There is little doubt that cockroaches contribute to the spread of several infections, mainly intestinal ones. Sometimes they may possibly be more important as mechanical vectors than house-flies. However, it remains difficult to assess their real importance as vectors because many of the pathogens which cockroaches carry can be transmitted in many other ways.

15.4   Control

Ensuring that neither food nor dirty kitchen utensils are left out overnight will help reduce the number of cockroaches, but if they are present in nearby houses, good hygiene in itself will not prevent them from entering houses.

Insecticidal spraying or dusting selected sites, such as cupboards, wardrobes, kitchen furniture, underneath sinks, stoves, refrigerators and dustbins, is recommended. Resistance to one or more of the organochlorine, organophosphate, carbamate and pyrethroid groups of insecticides has been reported, while resistance to all these groups is known in *Blatella germanica*. In the absence of resistance, sprays or dusts of pirimiphos-methyl, fenitrothion, malathion, diazinon, chlorpyrifos or bendiocarb can be used. Pyrethroids (e.g. permethrin, deltamethrin, cypermethrin, cyfluthrin) applied as sprays can produce spectacular results in both *flushing out* and killing cockroaches. Sprays or aerosols of insect growth regulators such as hydroprene, noviflumuron or pyriproxyfen, sometimes mixed with pyrethroids, can give effective cockroach control.

The residual action of sprays on painted and shiny surfaces may last only about 1–4 weeks, but they may be effective for several months on more porous surfaces. However, microencapsulated formulations have a longer residual life and are applied at lower dosages. The residual action of insecticidal dusts is less affected by the type of surface, but dusting is unsightly and so householders may object to it, and dusts should not be used in kitchens in case they contaminate food.

Commercial lacquers and varnishes containing residual insecticides such as 2% diazinon, 2% permethrin or 1% cypermethrin when painted on walls and other surfaces remain effective in killing cockroaches for several months.

Boric acid powder (borax) still remains a very safe and useful chemical, acting both as a contact insecticide and as a stomach poison.

Good control can be achieved when dinotefuran or imidacloprid (nicotinoids) or sulfluramid (sulfonamide) are added to baits, such as peanut butter, dog food and maltose, to which glycerol may be added to increase their attractiveness. Such poisonous baits are best placed in areas having large numbers of cockroaches. Alternatively cockroach pheromones can be placed in simple cardboard or sticky traps to entice cockroaches into them, after which they are either killed or prevented from escaping. However, baits by themselves will not eliminate cockroaches.

Further reading

Appel, A. G. and Smith, L. M. (2003) Biology and management of the smokeybrown cockroach. *Annual Review of Entomology*, **47**, 33–55.

Brenner, R. J., Barnes, K. C., Helm, R. M. and Williams, L. W. (1991) Modernized society and allergies to arthropods: risks and challenges to entomologists. *American Entomologist*, **37**, 143–55.

Burgess, N. R. H. and Chetwyn, K. N. (1981) Association of cockroaches with an outbreak of dysentery. *Transactions of the Royal Society of Tropical Medicine and Hygiene*, **75**, 332–3.

Cornwell, P. B. (1968). *The Cockroach. Volume 1: A Laboratory Insect and an Industrial Pest*. London: Hutchinson.

Cornwell, P. B. (1976) *The Cockroach. Volume 2: Insecticides and Cockroach Control*. London: Associated Business Programmes.

Elgderi, R. M., Ghenghesh, K. S. and Berbrash, N. (2006) Carriage by the German cockroach (*Blatella germanica*) of multiple-antibiotic-resistant bacteria that are potentially pathogenic to humans, in hospitals and households in Tripoli, Libya. *Annals of Tropical Medicine and Parasitology*, **100**, 55–62.

Gore, J. C. and Schal, C. (2007) Cockroach allergen biology and mitigation in the indoor environment. *Annual Review of Entomology*, **52**, 439–63.

Guthrie, D. M. and Tindall, A. R. (1968) *The Biology of the Cockroach*. London: Edward Arnold.

Pai, H. H., Chen, W. C. and Peng, C. F. (2005) Isolation of bacteria with antibiotic resistance from household cockroaches (*Periplaneta americana* and *Blatella germanica*). *Acta Tropica*, **93**, 259–65.

Petersen, R. K. D. and Shurdut, B. A. (1999) Human health risk from cockroaches and cockroach management: a risk analysis approach. *American Entomologist*, **45**, 142–8.

Roth, L. M. and Willis, E. R. (1960) The biotic associations of cockroaches. *Smithsonian Miscellaneous Collection*, **141**, 1–470.

Rust, M. K., Owens, J. M. and Reierson, D. A. (eds.) (1995) *Understanding and Controlling the German Cockroach*. New York, NY: Oxford University Press.

Rust, M. K., Reierson, D. A. and Hangsen, K. H. (1991) Control of American cockroaches (Dictyoptera: Blattidae) in sewers. *Journal of Medical Entomology*, **28**, 210–13.

Schal, C. and Hamilton, R. L. (1990) Integrated suppression of synanthropic cockroaches. *Annual Review of Entomology*, **35**, 521–51.

Stelmach, I., Jerzynska, J., Stelmach, W. *et al.* (2002) Cockroach allergy and exposure to cockroach allergen in Polish children with asthma. *Allergy*, **57**, 701–5.

16

Soft ticks (Argasidae)

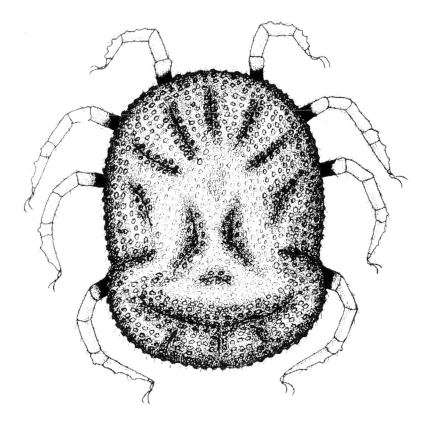

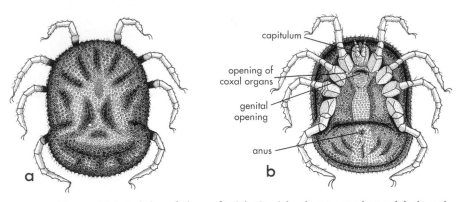

Figure 16.1 Adults of the soft tick *Ornithodoros moubata*: (a) dorsal view; (b) ventral view.

Ticks are not insects because adults have eight legs, not six as in adult insects. They are closely related to mites and spiders. Ticks are divided into two main families, the Argasidae (soft ticks) and the Ixodidae (hard ticks). A third family, Nuttalliellidae, contains just one species which is of no medical importance. Students sometimes find difficulty in distinguishing the very small immature stages of ticks from mites, but ticks differ from mites in having a *toothed* hypostome (Fig. 16.2), while adult ticks are also much *larger* than mites,

Soft ticks (Argasidae) have an almost worldwide distribution. There are some 185 species belonging to four genera, but the medically important soft ticks belong to the genus *Ornithodoros*. Species in this genus are found in many areas of the world including the Americas, Africa, Europe and Asia. The most important species is *Ornithodoros moubata*, a species in the *O. moubata* complex, which is a vector of tick-borne (endemic) relapsing fever (*Borrelia duttonii*). A few other species in the *O. moubata* complex are also of medical importance.

16.1   External morphology

Adult argasid ticks are *flattened* dorsoventrally, 8–13 mm long and usually roundish to oval in outline. The integument is wrinkled and usually has fine tubercles (mammillae) or granulations. There is *no scutum* (dorsal shield) as is found in ixodid (hard) ticks (Fig. 16.1a, Plate 21). The mouthparts, termed the *capitulum*, gnathosoma, or 'false head', are situated ventrally (Fig. 16.1b) and are *not visible* dorsally in the nymphs and adults. This one character separates adults and nymphal soft ticks from hard ticks (Ixodidae), but the larvae of *both* soft and hard ticks have the capitulum projecting forwards and clearly visible dorsally. The four-segmented *palps* are leg-like and the powerful cutting chelicerae have smooth, not

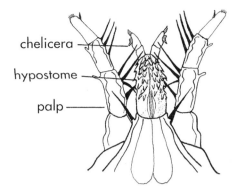

chelicera

hypostome

palp

Figure 16.2 Capitulum of an adult *Ornithodoros* species, showing leg-like palps and non-denticulate cheliceral sheaths.

denticulate, sheaths and teeth at their tips (Fig. 16.2). Both the *chelicerae* and the *hypostome* (Fig. 16.2), which has teeth arranged in several longitudinal rows, penetrate the host during feeding.

The four pairs of legs terminate in a pair of claws. *Coxal organs* ('glands') open between the bases of the coxae of the first and second pairs of legs (Fig. 16.1b), and are osmoregulatory in function.

Males and females look very similar and are usually difficult to separate, although blood-engorged females can be considerably larger than males because they ingest much more blood. However, because *both* sexes feed on blood and can consequently be disease vectors, it is not usually important to distinguish between them.

16.2   Internal anatomy

A brief account of the internal anatomy of a tick is necessary to understand the mechanisms of disease transmission.

During feeding, saliva, which usually contains powerful anticoagulants, is secreted by a pair of large grape-like salivary glands and flows down the mouthparts. The host's blood passes through the mouthparts and narrow oesophagus into the stomach (mid-gut), which has numerous branching diverticula. Side branches of the diverticula enable the adult tick to ingest large volumes of blood (about 6–12 times its own weight), causing great distension of the tick's body.

Argasid ticks have a pair of *coxal organs*, which although sometimes called coxal glands are not glandular but filter excess fluid and salts from ingested blood-meals. This fluid passes out through small openings located between the bases of the first two pairs of legs. When a soft tick is infected with tick-borne relapsing fever (*Borrelia duttonii*) many of the spirochaetes in the haemolymph enter the coxal organs and are passed out through their openings.

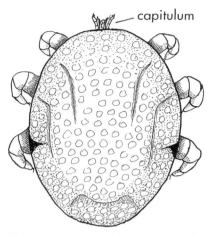

Figure 16.3 Larva of an *Ornithodoros* species, as an example of a typical soft tick showing the capitulum projecting in front of the body.

Coxal organs are present only in soft ticks, not in hard ticks.

Females of both soft and hard ticks have a peculiar structure called *Gene's organ* located in front of the mid-gut. During oviposition it is extruded from a small opening above the capitulum and secretes waxy waterproofing substances over the eggs, enabling them to withstand desiccation, immersion in water and other adverse environmental conditions.

16.3   Life cycle

A blood-meal is essential for egg production, and in argasids feeding is mainly nocturnal. Female ticks ingest large blood-meals, often increasing their weight 12-fold after feeding, while hard ticks ingest even more blood (p. 228). After each blood-meal female argasid ticks lay *several* (often 4–6) small egg batches, each comprising 15–100 spherical eggs. Occasionally an egg batch has as many as 300–500 eggs. Adult ticks can live for many years and females may lay thousands of eggs during their lifetime. Eggs are deposited in or near the resting places of adult ticks, such as in cracks and crevices in the walls, floors and furniture of houses, in mud, dust and debris, in rodent holes or in the more exposed resting or sleeping places of wild animals and birds.

Eggs usually hatch after 1–3 weeks, but because they have been coated during oviposition with a protective waxy secretion from Gene's organ (see above) they can remain viable for many months under adverse climatic conditions.

Both argasid and ixodid ticks have a *hemimetabolous* life cycle, that is eggs hatch to produce six-legged *larvae* which superficially resemble the adults, and which moult to produce eight-legged *nymphs* which resemble

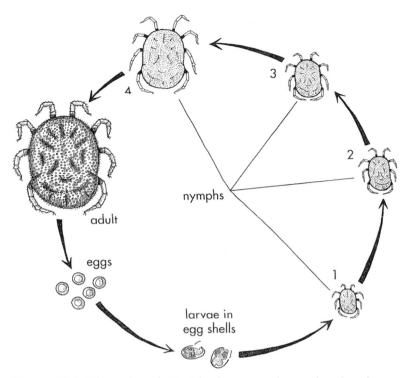

Figure 16.4 Life cycle of *Ornithodoros moubata*, showing larvae retained in eggshells and four nymphal stages.

even more closely the adults. In argasid ticks the six-legged larva (Fig. 16.3) is usually very active and searches for a host. In the larva the capitulum projects from the body and is visible from above. Blood-feeding may last 20–30 minutes, but in some species for several days, after which the engorged larva drops to the ground and after a few days moults to produce an eight-legged nymph. The nymph seeks a host and feeds for about 20–35 minutes before it falls to the ground. Argasid ticks usually have *four or five* nymphal instars (Fig. 16.4), but up to seven in some species. Each nymphal instar requires a blood-meal before it can proceed to the next stage. *Adults* usually feed on hosts for 20–35 minutes.

Larvae of *Ornithodoros moubata* differ from most other argasid ticks because they *do not* take blood-meals but remain within their eggshells after hatching, moulting to produce first-instar nymphs which crawl from the eggshells to seek blood-meals.

The duration of the *life cycle*, from egg hatching to adult, depends on the species of tick, temperature and availability of blood-meals, but in argasids is typically about 6–12 months. Adult ticks can *live* for many years, up to 12–20 years in the laboratory; the record for a tick is 25 years! In the absence of suitable hosts adults can survive up to 10 years without a blood-meal.

In argasid ticks mating usually occurs away from the host, such as on the ground or amongst vegetation.

Distribution of the larvae, nymphs and adults of argasid ticks is usually patchy and restricted to homes of their hosts. Such ticks are called nidicolous as opposed to the non-nidicolous ixodid ticks, which move further from homes of the hosts. Species which commonly feed on people, such as *O. moubata* in Africa, are found around human settlements, especially in village houses. They can also be found in livestock shelters, chicken sheds, animal burrows, and especially in dry areas in caves. However, in much of Africa argasid ticks appear to be becoming uncommon. This may be due to changes in lifestyle, in particular to the increased numbers of people sleeping on beds raised from the floor, which reduces the number of ticks feeding on them.

Because all nymphal instars and adults take blood, but nevertheless remain attached for only relatively short periods, many hosts, comprising different individuals and often different species, are fed upon during their life cycle. Argasid ticks are consequently referred to as *'many-host'* or 'multi-host' ticks.

16.4   Medical importance

16.4.1 Tick-borne relapsing fever

Soft ticks can inflict painful bites. They can sometimes cause tick paralysis, but this condition is more common following the prolonged feeding of hard ticks (p. 231).

Tick-borne relapsing fever is the only important disease transmitted to humans by soft ticks. The infection occurs throughout most of the tropics, subtropics and in many areas of the temperate region such as North America and Europe, but is absent from Australia and New Zealand.

There are about 15 species of *Borrelia*, mostly having different geographical distributions, that cause *Ornithodoros*-transmitted relapsing fevers. The most common is *B. duttonii*, found in sub-Saharan Africa and transmitted by *O. moubata*. In other geographical areas different ticks in the *O. moubata* complex transmit different species of *Borrelia*.

Spirochaetes ingested with a blood-meal multiply in the mid-gut, penetrate its wall and pass into the haemocoel, where they can be found after 24 hours. In the *haemocoel*, the spirochaetes multiply enormously and invade nearly all tissues and organs of the tick's body. After three days they infect the *salivary glands*, the *coxal organs* and *ovaries*. In *O. moubata* nymphs the salivary glands are more heavily infected than those of the adults. The coxal organs of nymphs are usually only lightly infected whereas those of the adults become heavily infected. When either nymphs or adults of *O. moubata* blood-feed saliva is injected into the bite, and spirochaetes can

Table 16.1. *Summary of principal features distinguishing soft and hard ticks*

Argasid ticks (Soft ticks)	Ixodid ticks (Hard ticks)
Morphology	
Scutum (shield) absent	Scutum (shield) on larvae, nymphs and adults. Females with small, and males with large scutums
Mouthparts (capitulum) not visible dorsally in nymphs and adults, but seen in larvae	Mouthparts (capitulum) visible dorsally in larvae, nymphs and adults
Palps leg-like; chelicerae have smooth sheaths	Palps club-shaped; chelicerae have denticulate sheaths
Coxal organs present	Coxal organs absent
Life cycle	
Eggs laid in several small batches of 15–100 eggs	Eggs laid in one large batch of many thousands of eggs
4–5 nymphal stages (8-legged)	Only one nymphal stage (8-legged)
Adults blood-feed rapidly, on hosts for only 20–35 minutes, but feed on several separate occasions	Adults feed slowly, on hosts for 1–4 weeks, but females feed only once
Multi-host ticks, usually about 6 hosts	Usually 2- or 3-host ticks
Ticks found mainly in or around homes of host, disperse little	Ticks attach to host for long time, hence can disperse considerable distances
Diseases	
Vectors of tick-borne relapsing fever	Vectors of tick-borne typhuses, Lyme disease and many viruses. Cause tick paralysis

be introduced by this route, especially by the nymphs. During feeding excess body fluids are filtered from the haemocoel by the coxal organs and in infected ticks, especially adults, the coxal fluids contain spirochaetes ingested with a previous blood-meal. These spirochaetes can enter the host through the puncture of the tick's bite or through intact skin. Humans can therefore become infected with *B. duttonii* by either the bite of *O. moubata* or the coxal fluids, or both.

In other *Ornithodoros* species the coxal organs tend to excrete excess fluids only when the ticks have left their host, and consequently transmission by these species is mainly by the tick's bite. In no species of *Ornithodoros* is infection spread by faeces.

The *tick* is usually regarded as the most important reservoir host, especially as there is ***transovarial*** transmission. That is ovaries of adult female ticks become infected with spirochaetes which are then passed to the eggs, so that the newly hatched larvae and all nymphal instars and adults, of both sexes, are infected. So, although nymphs and adults may not have fed on an infected person they can nevertheless transmit *B. duttonii* to other people. Such transovarial transmission can be continued for about three to four generations.

There may also be ***transstadial*** transmission. For example, a larva might become infected by feeding on an infected host and pass the spirochaetes to the nymphs and adults, or the infection might start with a nymph and be passed to subsequent nymphal instars and the adults. In all cases transovarial transmission can follow.

Although *B. duttonii*-transmitted relapsing fever is not a zoonosis, other tick-borne relapsing fevers, such as *Borrelia hermsii* transmitted by the tick *O. hermsi* in the Americas, are zoonoses, usually with rodents as reservoir hosts.

16.4.2 Q fever
Although Q fever is transmitted mainly by ixodid ticks, argasid ticks can also be vectors. See page 234 for an account.

16.4.3 Viruses
More than 100 arboviruses are transmitted by ticks, but only about 30 have been isolated from soft ticks, and very few infect people. Although soft ticks are not regarded as important vectors of arboviruses to humans, a new *Flavivirus* causing Alkurma haemorrhagic fever in Saudi Arabia is transmitted by *Ornithodoros savignyi*.

16.5   Control
Methods used for removing ticks from their hosts are described in Chapter 17 (p. 236).

Suitable repellents that can be applied to the skin include DEET, picaridin-based products, dibutyl phthalate or indalone. However, these repellents, especially DEET, are less effective against ticks than against mosquitoes. Alternatively clothing can be impregnated, or sprayed, with 0.5% permethrin.

Houses infested with argasid ticks, such as *Ornithodoros* species, can be sprayed with insecticides such as 0.5% carbaryl, 1% propoxur, 2% malathion, 0.5% chlorpyrifos, 0.5% diazinon or 1% pirimiphos-methyl, 0.3% permethrin, 0.3% deltamethrin or 1% cypermethrin. Care should be taken to spray floors and cracks and crevices in walls and furniture, and other sites where ticks may be resting. In houses where walls are uniformly plastered this usually reduces the numbers of ticks resting in them. When houses have been sprayed with residual insecticides for malaria control there is often a reduction in the numbers of *Ornithodoros*.

When applied to ticks or mites, insecticides are often called acaricides.

Further reading

Bowman, A. S. and Nuttall, P. A. (eds.) (2004) Ticks: biology, disease and control. *Parasitology* **129** (Suppl.), S1–S450.

Cunha, B. A. (ed.) (2001) *Tickborne Infectious Diseases: Diagnosis and Management*. New York, NY; Basel: Marcel Dekker.

Evans, G. O. (1992) *Principles of Acarology*. Wallingford: CAB International.

Goodman, J. L., Dennis, D. T. and Sonenshine, D. E. (eds.) (2005) *Tick-Borne Diseases of Humans*. Washington, DC: ASM Press.

Klompen, J. S. H., Black, W. C., Keirans, J. E. and Oliver J. H. (1996) Evolution of ticks. *Annual Review of Entomology*, **41**, 141–61.

Krantz, G. W. (1978) *A Manual of Acarology*, 2nd edn. Corvallis, OR: Oregon State University.

Lawrie, C. H., Uzcategui, N. Y., Gould, E. A. and Nuttall, P. A. (2004). Ixodid and argasid tick species and West Nile virus. *Emerging Infectious Diseases*, **10**, 653–7.

McCall, P. J. (2001) Tick-borne relapsing fever. In *The Encyclopedia of Arthropod-transmitted Infections of Man and Domesticated Animals*, ed. M. W. Service. Wallingford: CABI, pp. 513–16.

McDaniel, B. (1979) *How to Know the Mites and Ticks*. Dubuque, IA: W. C. Brown.

Obenchain, F. D. and Galun, R. (1982) *Physiology of Ticks*. Oxford: Pergamon Press.

Sauer, J. R. and Hair, J. A. (1986) *Morphology, Physiology, and Behavioral Ecology of Ticks*. Chichester: Ellis Horwood; New York, NY: Wiley.

Schuster, R. and Murphy, P. W. (eds.) (1991) *The Acari: Reproduction, Development and Life History Strategies*. London: Chapman and Hall.

Sonenshine, D. E. (1991) *Biology of Ticks, Volume 1*. New York, NY, and Oxford: Oxford University Press.

Sonenshine, D. E. (1993) *Biology of Ticks, Volume 2*. New York, NY, and Oxford: Oxford University Press.

See also references to hard ticks at the end of Chapter 17.

17

Hard ticks (Ixodidae)

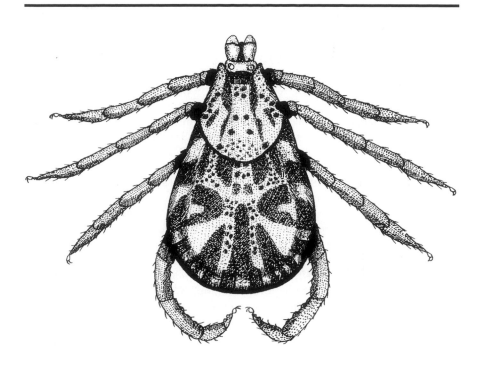

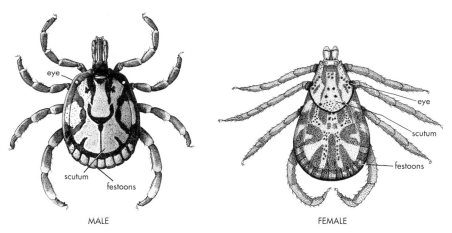

MALE   FEMALE

Figure 17.1 Adults of hard ticks: male *Amblyomma* (courtesy of the Natural History Museum, London) and female *Dermacentor*, showing sexual differences. A male ixodid has a large scutum while a female has a small scutum. Note the presence of festoons.

Hard ticks (Ixodidae) have a worldwide distribution, but are more common in temperate regions than soft ticks (Argasidae). There are 713 species of hard ticks belonging to 12 genera. Medically the more important genera are *Ixodes*, *Dermacentor*, *Amblyomma*, *Haemaphysalis*, *Rhipicephalus* and *Hyalomma*. Hard ticks are vectors of typhuses such as Rocky Mountain spotted fever (*Rickettsia rickettsii*) and Mediterranean spotted fever (*R. conorii*), and Q fever (*Coxiella burnetii*). Many arboviruses, including tick-borne encephalitis, Omsk haemorrhagic fever, Kyasanur Forest disease, Crimean–Congo haemorrhagic fever and Colorado tick fever, are transmitted by hard ticks. They also transmit tularaemia (*Francisella tularensis*), and cause tick paralysis.

17.1   External morphology

Adult hard ticks are *flattened* dorsoventrally, oval in shape and about 2–23 mm long, size depending on species and whether they are unfed or fully engorged with blood. Females are usually bigger than males, and because they take larger blood-meals they enlarge much more than males during feeding.

The *capitulum* or 'false head' projects forwards from the body and is visible from above (Fig. 17.1, Plate 22), thus distinguishing adult hard (ixodid) ticks from soft (argasid) ticks (Fig. 16.1). Also in hard ticks the *palps* are swollen and club-shaped (Fig. 17.2) rather than leg-shaped as in soft ticks, and unlike soft ticks the cheliceral sheaths are covered with very small denticles. As in argasid ticks both the *hypostome* and *chelicerae* penetrate the host during feeding. In hard, but not soft, ticks a cement-like substance

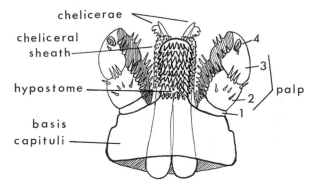

Figure 17.2 Capitulum of an adult ixodid tick, showing club-shaped palps with minute fourth segment, and denticulate cheliceral sheaths.

from the salivary glands 'glues' the mouthparts firmly into the host, and ensures continued attachment during their long feeding times (days to weeks).

The posterior margin of the body in species of *Dermacentor, Rhipicephalus* and *Haemaphysalis* has a number of rectangular indentations called *festoons*. However, in fully engorged females these indentations may be difficult to see due to the body's distension with blood.

Hard ticks have a dorsal plate called a dorsal shield or *scutum*, which is absent in soft ticks. In *males* the scutum is large and covers almost the entire dorsal surface of the body (Fig. 17.1), whereas in *females* it is much smaller and restricted to the anterior part of the body (Fig. 17.1). In fully fed females the scutum may be difficult to see because it appears small in relation to the enlarged body and becomes pushed forwards so that it is almost vertical in position. In both sexes of *Dermacentor, Amblyomma* and some *Rhipicephalus* species the scutum has so-called enamelled coloured areas, and such ticks are described as *ornate* species. The scutum provides a method of immediately recognizing hard ticks and also of distinguishing the sexes, although medically this may not be very important because *both* sexes take blood-meals and are therefore potential disease vectors. In the larval and nymphal stages the scutum is small in both sexes.

There are four pairs of legs, with each leg ending in a pair of claws. Coxal organs are absent in hard ticks. The internal organs are basically as encountered in argasid ticks (pp. 217–18).

17.2   Life cycle

Both ixodid (hard) and argasid (soft) ticks have *hemimetabolous* life cycles, that is there is incomplete metamorphosis involving a *larval* and *nymphal* stage. There are, however, important differences between the life cycles and ecology of hard and soft ticks. *Adult* ixodid ticks remain attached to their hosts for long periods as blood-feeding often lasts 1–4 weeks. After

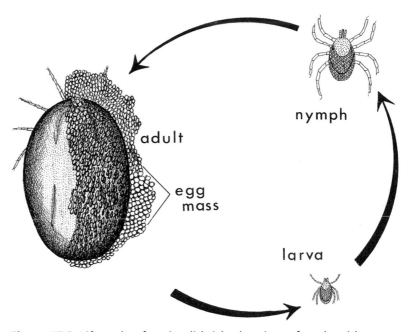

Figure 17.3 Life cycle of an ixodid tick, showing a female with a very large egg mass and a single nymphal stage.

feeding the enormously engorged tick drops from the host to the ground and shelters under leaves, stones, amongst surface roots of grasses and shrubs, or buries itself in the surface soil. These are termed non-nidicolous ticks, in contrast to the nidicolous argasid ticks, which live in hosts' burrows and homes. Time taken for females to digest their blood-meal and commence laying eggs varies according to species and environmental conditions, especially temperature. Sometimes oviposition begins 3–6 days after the female drops from the host, but egg-laying may not begin until several weeks or occasionally months after the end of feeding. *Thousands* (1000–10 000) of small spherical eggs are laid in a gelatinous mass which is formed in front and on top of the tick's scutum (Fig. 17.3). A few species lay as many as 20 000 eggs, and the egg mass may become larger than the ovipositing female. Oviposition can last 10 days, or extend over a month or more. As in argasid ticks the eggs are coated with a waxy secretion produced by *Gene's organ*, which in ixodid ticks also helps transfer eggs from the genital opening to the scutum. The female ixodid tick lays *only* one batch of eggs, after which she dies.

After 10–20 days to a few months six-legged larvae hatch from the eggs. The larvae are minute (0.5–1.5 mm long), and are sometimes called *seed ticks*. On cursory examination they superficially resemble larval mites, but the presence of a *toothed* hypostome immediately identifies them as ticks. Newly emerged larvae remain inactive for a few days, after which they climb up vegetation and wait for passing hosts. When suitable hosts occur they extend their front legs in the air and cling to the hairs or feathers of passing hosts. Such host-seeking behaviour exhibited by the larvae, and also by the nymphs and adults, is called *questing*. Once on a host the chelicerae and hypostome are inserted deep into the skin and the larvae commence blood-feeding. *Larvae* remain on their hosts for about 3–7 days before dropping to the ground and sheltering amongst vegetation or under stones. They normally take a few days to digest their blood-meals, but in cooler weather digestion may extend over several weeks. After blood has been digested the larvae remain inactive for a further few days before they moult to become nymphs.

Newly formed eight-legged *nymphs* climb up vegetation and behave similarly to the larvae (that is *questing*) in seeking a host. *Blood-feeding* lasts 5–10 days, after which fully engorged nymphs drop to the ground and shelter under stones or amongst vegetation. They remain quiescent for a few weeks while the blood-meal is digested, after which the nymphs moult to produce male or female ixodid ticks. There is only *one* nymphal stage in the life cycle of ixodid ticks (Fig. 17.3), whereas argasid ticks have several nymphal stages.

Newly formed adults remain more or less inactive for about a week, after which they climb vegetation and start questing for passing hosts. *Adult* female ticks can take enormous blood-meals, ingesting 1–8 ml of blood. Although much of the fluid of this is excreted the weight of an unfed female can nevertheless increase some 200 times after blood-feeding! In contrast male ticks ingest much less blood. Adults shelter under stones and surface vegetation, and amongst roots of plants. Except for *Ixodes* species, mating in ixodid ticks occurs on the host, and males may remain on a host for several weeks or months, mating with several females.

17.3   Behaviour and habits

Many species of ixodid ticks are more or less host-specific. For example some species feed almost exclusively on birds, others on reptiles, or on certain types of mammals such as bats, canids or bovids, but other species feed on almost any available hosts, including people. Diversity of host species may increase the likelihood of disease transmission. *Larvae and nymphs* of many, but not all, ticks seem to favour feeding on small animals such as rodents, cats and dogs, and ground-inhabiting birds, whereas *adults* prefer to feed on cattle, deer, horses and a variety of large and wild mammals. *Humans* are mainly parasitized by larvae and nymphs of hard ticks, more rarely by adults.

The *life cycle* of ixodid ticks commonly takes 2–4 years, depending on species and availability of hosts, but may be longer due to lack of hosts. In temperate regions development may also be prolonged or cease temporarily during winter months. In warm countries development and breeding often continue throughout the year; nevertheless there may be seasonal fluctuations in development rates. *Adult* ticks usually live for 2–3 years, although some can live up to 7 years, and adults may be able to survive 2–3 years of starvation. Although ticks can tolerate considerable variations in temperature and humidity, most species are absent from very dry or very wet areas, although certain *Hyalomma* species occur in arid areas. Microclimatic conditions at soil level are greatly influenced by the amount and type of ground vegetation, and consequently the distribution of tick species is often closely associated with different types of vegetation.

Because immature and adult ixodid ticks remain on hosts much longer than do argasid ticks they may be carried many kilometres by their hosts, or even across continents by migrating birds.

17.3.1 Three-host ticks
The life cycle described for hard ticks refers to a three-host tick. That is a different individual host, which may be the same or different species, is parasitized by the larva, nymph and adult, and moulting occurs on the ground. About 95% of ixodid ticks have this type of life cycle, and medically important species of three-host ticks are found in the genera *Ixodes, Dermacentor, Rhipicephalus, Haemaphysalis* and *Amblyomma*. Ticks which feed on three hosts are more likely to become infected with pathogens and be potential vectors than species feeding on just one or two hosts.

17.3.2 Two-host ticks
Larvae of some *Hyalomma* and *Rhipicephalus* species remain on the host after blood-feeding and moult to produce nymphs which then feed on the same host. The engorged nymphs then drop off, moult and the resultant adults feed on a different host. This a two-host tick life cycle.

17.3.3 One-host ticks
In a few ticks, such as *Boophilus* species, the larva, nymph and adult all feed on the same host and moulting also takes places on that host. The only stage that leaves the host is the blood-engorged female tick, which drops to the ground to lay eggs. One-host ticks are less likely to acquire infections with pathogens than ticks which feed on several hosts, and the only method by which infection can spread from one host to another is by transovarial transmission. One-host ticks are of little or no medical importance, but certain species of *Boophilus* are important vectors of animal infections.

See Table 16.1 (p. 221) for a summary of the differences between hard and soft ticks.

Table 17.1. *Some infections transmitted to humans by hard ticks*

Disease	Infective agent	Principal tick vectors	Main reservoir hosts excluding ticks
Tick-borne encephalitis	*Flavivirus*	*Ixodes ricinus* *I. persulcatus*	Rodents, insectivores
Omsk haemorrhagic fever	*Flavivirus*	*Dermacentor reticulatus*	Muskrats, water voles
Kyasanur Forest disease	*Flavivirus*	*Haemaphysalis spinigera* *H. turturis*	Monkeys, possibly shrews
Crimean–Congo haemorrhagic fever	*Nairovirus*	*Hyalomma marginatum* complex	Hares, cattle, goats
Colorado tick fever	*Coltivirus*	*Dermacentor andersoni*	Many rodent species
Rocky Mountain spotted fever	*Rickettsia rickettsii*	*Dermacentor, Amblyomma* and *Rhipicephalus* species	Many rodent species
Mediterranean spotted fever	*Rickettsia conorii*	*Rhipicephalus sanguineus*	Rodents, dogs
African tick-bite fever	*Rickettsia africae*	*Amblyomma* species	Rodents, possibly cattle
Q fever	*Coxiella burnetii*	Many ixodid species	Sheep, goats, cattle, possibly rodents
Human monocytic ehrlichiosis	*Ehrlichia chaffeensis*	*Amblyomma americanum*	Deer, rodents
Lyme disease	*Borrelia burgdorferi*	*Ixodes ricinus* *I. scapularis* *I. pacificus*	Rodents, birds
Tularaemia	*Francisella tularensis*	Many ixodid species	Rabbits, hares, deer, beavers
Tick paralysis	Tick toxins	Mainly *Ixodes* and *Dermacentor* species	Not applicable, as not caused by any pathogen

17.4 Medical importance
Table 17.1 very briefly summarizes some of the infections transmitted to humans by hard ticks. More details are given below under the various headings.

17.4.1 Tick paralysis
Female hard ticks, mainly species of *Dermacentor* and *Ixodes*, can cause tick paralysis. Human cases have been reported from North and South America, Europe, Asia, Australia and South Africa. The condition also affects pets and domesticated animals. Symptoms appear 4–7 days after a tick, usually a female, has commenced feeding. There is an acute ascending paralysis affecting firstly the legs, resulting in the patient being unable walk or stand, and later the arms cannot be moved and there follows difficulty in speaking, swallowing and breathing. Symptoms are painless and there is rarely any rise in the patient's temperature. Tick paralysis can be confused with paralysis due to poliomyelitis and certain other paralytic infections. Young children, especially those up to two years, are most severely affected. Death in animals, and in rare cases humans, can result from respiratory failure. Removal of ticks can result in complete recovery after 48 hours, but in severe cases recovery may take a few days, or even up to about six weeks.

Tick paralysis is not caused by pathogens but by toxins in the female tick's saliva which are continually being pumped into the host during the tick's long feeding period. Different species of ticks and also different populations of the same species may vary markedly in their ability to produce tick paralysis in humans and animals.

17.4.2 Arboviruses
More than 120 arboviruses have been recovered from ticks, but the important tick-borne viral diseases of humans are spread by hard ticks. All arboviruses are transmitted by the tick's bite, and *transovarial* transmission usually occurs.

Tick-borne encephalitis (TBE) (*Flavivirus*)
First described in 1932 as Russian spring–summer encephalitis (RSSE), and then discovered after World War II (1939–45) in central Europe and called central European encephalitis (CEE). These two infections are now considered by many to be the same, and it is called tick-borne encephalitis (TBE). It is widespread in Europe (except the UK, Benelux countries and the Iberian peninsula), Siberia, northern Asia, China and Japan.

In Siberia eastwards the main vector is *Ixodes persulcatus*, while in Europe it is *I. ricinus*. After multiplication in the tick, virus accumulates in the salivary glands, and infection is through the tick's bite. Small rodents (e.g. voles) and insectivores (e.g. hedgehogs) in addition to ticks are *reservoir hosts*. There is *transstadial* and *transovarial* transmission. Humans are not part of the natural transmission cycle but become accidentally infested with ticks.

TBE virus accumulates in the mammary glands of goats, sheep and cows, and people may become infected by drinking infected unpasteurized milk or eating cheese.

Omsk haemorrhagic fever (OHF) (*Flavivirus*)

The virus causing OHF is antigenically very similar to viruses causing TBE and Kyasanur Forest disease (KFD), and clinical symptoms are rather similar to those caused by these other viruses. OHF occurs in Siberia, such as in the Omsk region. The primary vector is *Dermacentor reticulatus* (formerly called *D. pictus*), which feeds on rodents, especially *muskrats*, which are *amplifying hosts*; while water voles are also possible amplifying and reservoir hosts. Other important vectors are *D. marginatus* and *Ixodes persulcatus*. Infections acquired from animal hosts are transmitted *transstadially* to nymphs or adults. *Transovarial* transmission also occurs. Muskrat hunters are particularly liable to come into contact with infected ticks, and the disease can also be passed directly to them by the animals' urine and faeces. Infection can also be through drinking milk of goats or sheep.

Kyasanur Forest disease (KFD) (*Flavivirus*)

KFD was first recognized in 1957 when monkeys were dying in Kyasanur Forest in Karnataka State of southern India and people were also becoming ill and dying. The disease is now found in about 5000 km² in and around Kyasanur Forest and is associated with movements of people into forests, cattle grazing at the forest edge and deforestation for food crops, activities which expose people to ticks. The main vectors are species of *Haemaphysalis*, especially *H. spinigera*, which transmits the virus to humans, while *H. turturis* maintains animal transmission.

Larval ticks feed on birds and small forest rodents, whereas the nymphal stages feed mainly on monkeys and humans. *Monkeys*, and possibly shrews, seem to be the principal *amplifying hosts* and reservoir hosts. Larger mammals such as deer, bison and cattle brought to the forest edge to graze serve as hosts for adult ticks and help maintain large tick populations, but are not reservoir hosts of viral infection. There is *transstadial* and probably *transovarial* transmission in the tick vectors. The epidemiology of KFD is particularly interesting because it shows how changes in people's behaviour, such as deforestation and agricultural development, can lead to changing ecology and disease outbreaks in the human population.

Crimean–Congo haemorrhagic fever (CCHF) (*Nairovirus*)

CCHF virus is recorded from many countries in central and southern Europe, the Middle East, Russia, Pakistan, India, China, Madagascar and in Africa from Mauritania to Ethiopia down to South Africa. After dengue viruses CCHF virus is one of the most widely distributed arboviruses, with human infections known from about 30 countries and virus isolations obtained from ticks in another 10 countries. The disease is typically enzootic in savannah, steppe and semidesert areas. Transmission is mainly by *Hyalomma* species, such as *H. marginatum marginatum*, but in Africa *H. marginatum rufipes* is the vector. Larval and nymphal ticks feed on birds and small mammals, while adults feed on larger mammals including humans. Hares, cattle and goats are *amplyfying hosts* and likely reservoir hosts. Although *birds* are not reservoir hosts they can spread infected ticks around the world; for example some 5 billion birds fly annually from Europe to Africa, and about half return.

Transmission is by tick bite and possibly by crushing infected ticks, or by accidental contamination from infected blood during sheep-shearing. *Ticks* are regarded as important reservoirs of infection, especially as they can survive starvation for at least 800 days! There is *venereal* transmission, i.e. virus is transmitted from infected male ticks to uninfected females during mating, and then *transovarial* transmission from infected females to their progeny followed by *transstadial transmission*.

Colorado tick fever (CTF) (*Coltivirus*)
CTF occurs in the Rocky Mountain states and South Dakota in the USA and in western Canada. The principal vector is *Dermacentor andersoni*. Larvae and nymphs feed on small mammals such as rabbits, ground squirrels, chipmunks and woodrats, which together with ticks are the main *reservoir hosts* of infection. Adult ticks, and sometimes nymphs, feed on larger mammals, such as deer, cattle and people. There is both *transstadial* and *transovarial* transmission.

17.4.3 Rickettsiae
Tick-borne typhuses have an almost worldwide distribution and are caused by 13 species of *Rickettsia*. Ticks are usually regarded as the main *reservoirs* of infection although rodents and other mammals may sometimes by reservoir hosts. There is usually *transovarial* transmission and often *transstadial* transmission. The more important tick-borne typhuses are described briefly below.

Rocky Mountain spotted fever (RMSF)
RMSF, also known as Mexican spotted fever, São Paulo spotted fever, American tick-borne typhus and by several other local names, occurs in North, Central and South America. The causative agent is *Rickettsia rickettsii*. The principal vector in western America is *Dermacentor andersoni* and in eastern USA *D. variabilis*, and recently *Rhipicephalus sanguineus* has been found to be a vector in Arizona. In South America *Amblyomma cajennense*

is the main vector, and this species and *Rhipicephalus sanguineus* are the important vectors in Central America. RMSF is a *zoonosis*, with ground squirrels, chipmunks and other small rodents acting as *reservoir hosts* and/or *amplifying hosts*, although the tick itself is considered the main reservoir of infection, especially as infections can persist in overwintering ticks. Dogs are not reservoir hosts or amplifying hosts, but they can transport infected ticks to human habitations, where they may become dislodged and attach to people.

Transmission is normally through the bite of any stage in the life cycle of the tick. An infective tick, however, must remain feeding on a host for at least two hours before sufficient rickettsiae are injected into the host for the host to become infected. Consequently, early tick removal may prevent transmission. There is both *transstadial* and *transovarial* transmission.

Mediterranean spotted fever

Also known as boutonneuse fever, Marseilles fever, South African tick typhus, Kenyan tick typhus, Indian tick typhus and Crimean tick typhus. The infective agent is *Rickettsia conorii*. It occurs in the Mediterranean littoral region, Israel, Portugal, Sicily, eastern Russia, India, sub-Saharan Africa and rather surprisingly also in Uruguay. The principal vector is *Rhipicephalus sanguineus*, the dog tick. Transmission is by the tick's bite, and both *transstadial* and *transovarial* transmission occur. Ticks, various rodents and, in contrast to RMSF, dogs can be *reservoir hosts*. Infection can also occur if infected ticks are *crushed* and the rickettsiae rubbed into abrasions or the eyes.

African tick-bite fever

Initially confused with typhus caused by *R. conorii*, but in 1992 the causative agent was named *R. africae*. This form of typhus is common throughout most of sub-Saharan Africa, and also occurs in the West Indies. In both regions vectors are *Amblyomma* species. Rodents and possibly cattle are *reservoir hosts*.

Miscellaneous tick-borne typhuses

These include Siberian tick typhus (*R. sibirica*), Queensland tick typhus (*R. australis*) and Japanese or Oriental tick typhus (*R. japonica*) as well as unnamed diseases caused by other *Rickettsia* species.

Q fever

Q fever is a rickettsial zoonotic disease caused by *Coxiella burnetii*. It has a worldwide distribution and is primarily an infection of rodents, other small mammals and domestic livestock. It can be transmitted to people by inhalation of aerolized rickettsia, possibly by drinking contaminated milk, and occasionally by bites of argasid ticks, but mainly ixodid ones.

Sheep, goats and cattle, and possibly rodents, are *reservoir hosts*, while ticks are probably important in maintaining infections in wild animals and in transmission to domesticated ones. *Transovarial* and *transstadial* transmission occurs.

17.4.4 Ehrlichiosis

Species of *Ehrlichia* and *Anaplasma* infect dogs and ruminants, and some species are zoonotic and infect humans. One species, *E. chaffeensis*, causes human monocytic ehrlichiosis (HME), while *Anaplasma phagocytophilum* (= *E. phagocytophila*) and *E. ewingii* parasitize granulocytes and cause human granulocytic ehrlichiosis (HGE). Transmission of all three species is by bites of hard ticks such as *Amblyomma* and *Ixodes* species, while rodents and deer appear to be the main *reservoir hosts*. *Transstadial* transmission occurs, and possibly also transovarial. Ehrlichiosis is widespread in Europe and the USA, where infections seems to be increasing in prevalence, but ehrlichiosis has also been reported from Africa and Venezuela.

17.4.5 Spirochaetes
Lyme disease

Lyme disease (also called Lyme borreliosis or erythema migrans) was first recognized in 1975 in the town of Old Lyme, Connecticut, USA. It is caused by the spirochaete *Borrelia burgdorferi*, one of 11 species within the *B. burgdorferi* complex. Here the parasite will be referred to as *B. burgdorferi*, although in some cases of Lyme disease the infectious agent may be another species, such as *B. afzellii*.

Lyme disease occurs in at least 26 European countries, Asia, China, Japan, the USA, Canada, Africa and possibly Australia. In Europe transmission is by the bite of *Ixodes ricinus* and in Eurasia by *I. persulcatus*. In the eastern USA the vector is *Ixodes scapularis* (= *I. dammini*) whereas in western areas *I. pacificus* is the principal vector. There is both *transstadial* and *transovarial* transmission. Lyme disease is the most commonly reported arthropod-transmitted infection in North America and Europe. In the USA a peak of 23 763 cases were reported in 2002, while in 2004 the number was 19 723.

Lyme disease is a *zoonosis*. More than 100 animal species have been identified as being infected, but rodents and birds are the most important *reservoir hosts*. Deer, although supporting large populations of vector ticks, are not reservoir hosts.

The ecology of Lyme disease and reasons for its increased prevalence and the extension of its geographical range are both complex and interesting. One explanation is that in both North America and Europe people are spending more time in rural areas where there can be large infestations of deer ticks. Moreover, people are increasingly building homes near to recently cleared land adjacent to forests, and this increases their exposure to ticks. Greater awareness and more extensive serological testing of people may also be partly responsible for the increased numbers of reported cases.

17.4.6 Tularaemia

Tularaemia is a bacterial disease caused by *Francisella tularensis*. It occurs throughout the northern hemisphere, being most common in the USA, where 150–300 human cases occur each year, and in Europe. It infects mainly rabbits and hares, but also small rodents, beavers and deer, all of which can be *reservoir hosts*. The infection is spread by a variety of direct contact methods such as handling infected live animals or carcasses, drinking contaminated water, eating raw and uncooked meats, and also by the bites of various hard ticks. In Europe the main vectors are *Ixodes ricinus* and *Dermacentor* species. The tabanid fly *Chrysops discalis* has also been identfied as a vector in North America.

17.5   Control

Personal protection methods include application of repellents, as for soft ticks (Chapter 16).

Many methods have been advocated for removal of ticks from their hosts, including coating them with vaseline, medicinal paraffin or nail varnish. But it may be several hours before such ticks withdraw their mouthparts and this is usually unacceptable, especially as *rapid removal* of ticks often reduces the chances of disease transmission. With soft (argasid) ticks relatively rapid removal is sometimes achieved by dabbing them with chloroform or some other anaesthetic or spraying them with 0.5% permethrin. But this method rarely works with hard ticks because they are attached to their hosts with a type of 'saliva cement' which prevents rapid withdrawl of their mouthparts. The recommended procedure is to grasp the tick as close as possible to the host's skin with blunt forceps and slowly pull the tick out. The mouthparts (capitulum) may remain imbedded in the skin and should be removed if possible, and then an antiseptic applied.

Many species of hard ticks transmit livestock infections, and worldwide regular dipping of sheep and cattle in acaricidal baths, or spraying them with insecticides (acaricides) is widely undertaken. Such intensive use of insecticides has resulted in ticks becoming resistant to many types of insecticides; nevertheless insecticides are still needed to control medically important ticks. For example, dogs can be treated with sprays, lotions or dusts containing permethrin, resmethrin, carbaryl, propoxur or chlorpyrifos to kill their ticks, which may otherwise attach to humans. Commercially available 'spot-on' treatments using permethrin, fenthion, imidacloprid, amitraz or fipronil are also available to kill dog ticks.

Floors of houses, porches, verandahs and other sites where infested pets sleep should be sprayed with insecticides, as outlined on page 222 for control of endophilic soft ticks.

Ticks in gardens, yards and nearby fields can be controlled by spraying 0.5% diazinon, 0.5% carbaryl, 0.1% propoxur, 0.25% permethrin or other pyrethroids. Pellet formulations are best when there is dense vegetation because they more easily penetrate ground cover to reach the microhabitats harbouring ticks; a single application may be effective for 6–8 weeks. ULV (ultra-low-volume) spraying with insecticides reduces dosage rates, but there may be environmental objections to such blanket-type spraying of vegetation.

Further reading

Camicas, J.-L., Hervy, J.-P., Adam, F. and Morel, P.-C. (1998) *The Ticks of the World* (Acarida, Ixodida): *Nomenclature, Described Stages, Hosts, Distribution.* Paris: Editions de l'ORSTOM.

CDC (2005) Tularemia transmitted by insect bites: Wyoming 2001–2003. *MMWR Weekly*, **54** (07), 170–3.

Dumler, J. S. and Walker, D. H. (2005) Rocky mountain spotted fever: changing ecology and persisting virulence. *New England Journal of Medicine*, **353**, 551–3.

Gammons, M. and Salam, G. (2002) Tick removal. *American Family Physician*, **66**, 643–5.

Gothe, R., Kunze, K. and Hoogstraal, H. (1979) The mechanisms of pathogenicity in the tick paralyses. *Journal of Medical Entomology*, **16**, 357–69.

Gray, J. S., Kahl. O., Lane, R. S. and Stanek, G. (2002) *Lyme Borreliosis: Biology, Epidemiology and Control.* Wallingford: CABI.

Hoogstraal, H. (1966) Ticks in relation to human diseases caused by viruses. *Annual Review of Entomology*, **11**, 261–308.

Hoogstraal, H. (1967) Ticks in relation to human diseases caused by *Rickettsia* species. *Annual Review of Entomology*, **12**, 377–420.

Hoogstraal, H.(1981) Changing patterns of tickborne diseases in modern society. *Annual Review of Entomology*, **26**, 75–99.

Kisinza, W. N., McCall, P. J., Mitani, H., Talbert, A. and Fukunga, M. (2003) A newly identified tick-borne *Borrellia* species and relapsing fever in Tanzania. *Lancet*, **362**, 1283–4.

Lane, R. S., Piesman, J. and Burgdorfer, W. (1991) Lyme borreliosis: relation of its causative agent to its vectors and hosts in North America and Europe. *Annual Review of Entomology*, **36**, 587–609.

Needham, G. R. and Teel, P. D. (1991) Off-host physiological ecology of ixodid ticks. *Annual Review of Entomology*, **36**, 313–52.

Parola, P. and Raoult, D. (2001) Tick-borne typhuses. In *The Encyclopedia of Arthropod-transmitted Infections of Man and Domesticated Animals*, ed. M. W. Service. Wallingford: CABI, pp. 516–24.

Schultz, G. W., Robbins, R. G., and Hill, D. W. (2005) *Interactive Program for Teaching Tick Morphology.* Version 1, CD ROM. Defense Pest Management Information Analysis Center, Armed Forces Pest Management Board, Walter Reed Army Medical Center, Washington, DC. www.afpmb.org/bulletin/vol25/tickcd.htm.

Sonenshine, D. E. (2006) Tick pheromones and their use in tick control. *Annual Review of Entomology,* **51**, 557–80.

Sonenshine, D. E., Lane, R. S. and Nicholson, W. L. (2002) Ticks (Ixodida). In *Medical and Veterinary Entomology,* ed. G. Mullen and L. Durden. Amsterdam: Academic Press, pp. 517–58.

Sonenshine, D. E. and Mather, T. N. (eds.) (1994) *Ecological Dynamics of Tick-Borne Zoonoses.* New York, NY: Oxford University Press.

Steere, A., Coburn, J., Glickstein, L. (2005). Lyme borreliosis. In *Tick-Borne Diseases of Humans,* ed. J. L. Goodman, D. T. Dennis and D. E. Sonenshine. Washington, DC: ASM Press.

See also references to soft ticks at the end of Chapter 16.

18

Scabies mites (Sarcoptidae)

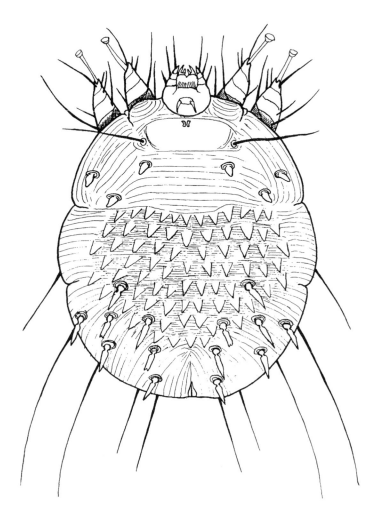

Adult mites, like ticks, have eight legs and therefore are not insects. They can be distinguished from ticks by the *absence* of teeth on the hypostome of the mouthparts and in having *setae* (bristles) on the body as well as the legs. But the principal medically important species (scabies mite, scrub typhus mite, house-dust mite and follicle mite) can most readily be recognized by their characteristic shapes.

Sarcoptes scabiei, the scabies or itch mite, occurs on people worldwide. Morphologically they are inseparable from *S. scabiei* infesting wild and domesticated animals, including dogs, horses and pigs. Mites on such animals are considered to be the same species as those infecting people but physiologically adapted for life on non-human hosts. In animals they cause the condition known as mange. Mites living on animals very rarely infect humans, but if they do the infection can persist for several weeks.

Scabies mites are not vectors of any disease but cause conditions known as scabies, acariasis, and crusted or Norwegian scabies.

18.1   External morphology

The female mite (0.30–0.45 mm) is just visible without the aid of a hand lens. It is pale and disc-shaped. Dorsally the mite is covered with numerous small *peg-like spines* and a few bristles (setae), and both dorsally and ventrally there are series of lines across the body giving the mite a striated appearance (Fig. 18.1, Plate 23). Adults have four pairs of short and cylindrical legs divided into five ring-like segments. The first two pairs of legs end in short stalks called *pedicels* which terminate in thin-walled roundish structures often termed 'suckers'. In *females* the posterior two pairs of legs do not have 'suckers' but end in long and very conspicuous bristles. There is no distinct head, but the short and fat palps and pincer-like *chelicerae* of the mouthparts protrude anteriorly from the body.

Adult male scabies mites are only 0.20–0.25 mm long, and apart from their small size may also be distinguished from females by the presence of 'suckers' on the last pair of legs (Fig. 18.2).

18.2   Life cycle

Scabies was for a long time associated with poverty, overcrowding and poor hygiene, but people from all socioeconomic backgrounds can become infested with mites.

Female mites dig and eat their way into the surface layers of the skin – the stratum corneum – selecting places where the skin is thin and wrinkled, such as between the fingers and on the wrists, elbows, feet, penis, scrotum, buttocks and axillae. The majority (63%) of mites are found on the hands and wrists, and about 11% on the elbows. In women mites may be found burrowing beneath and around the breasts and nipples. Occasionally mites are found on the face and scalp, especially in the postauricular fold, and on other parts of the body. The greatest number of mites on children up

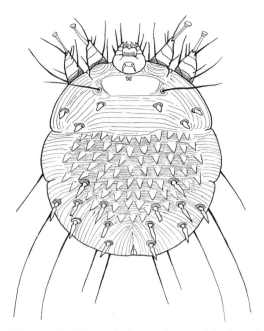

Figure 18.1 Dorsal view of an adult female scabies mite (*Sarcoptes scabiei*).

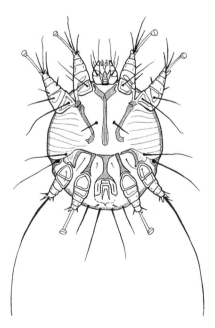

Figure 18.2 Ventral view of the smaller adult male scabies mite, showing 'suckers' on the hind-legs (absent on the hind-legs of females).

to a year old are often found on the feet. When females burrow into the skin they excavate winding tunnels at the rate of about 0.5–5 mm per day, which are seen on the skin as very thin twisting lines measuring a few millimetres to 1–2 cm. Mites feed on liquids oozing from dermal cells they have chewed. A female mite lays 1–3 eggs a day in her tunnel.

Eggs hatch after 3–4 days, and small six-legged *larvae* emerge which look like miniature adults. They crawl out of the tunnels onto the surface of the skin where a large number die, but a few succeed in either burrowing into the stratum corneum or entering a hair follicle to produce not a tunnel but a small pocket called a *'moulting pocket'*. After 3–4 days the larva moults in the pocket to produce an eight-legged *nymph* (protonymph) which after 2–4 days moults to produce a second nymph (tritonymph). In mites destined to become females this second nymphal stage (tritonymph) is considerably larger than the nymph of a male. After a further 2–3 days the second nymphal stage (tritonymph) moults to become either a male or a female adult. After mating the female adult increases in size and is sometimes referred to as an ovigerous or mature female. Only after fertilization does the female commence to burrow through the skin, and after 3–5 days begins to lay eggs in her tunnel. Female mites rarely leave their tunnels, but contrary to previous beliefs dislodged healthy female mites can sometimes be found on bed linen and on clothing.

Adult male mites are about half the size of ovigerous females and can be found either in very short burrows (usually less than 1 mm) or in small pockets in the skin. However, they probably spend most of their life wandering around on the surface of the skin looking for females.

The *life cycle* from egg to adult usually takes 10–14 days. Female mites may live for 4–6 weeks on humans, but away from their hosts for only 2–5 days.

Scabies is a contagious complaint which is transmitted only by close contact. It is therefore a family disease spreading amongst those living in close association, especially when they share the same bed. Scabies can be spread between courting couples who are habitually holding hands. It appears that the actual *transfer* of mites from person to person takes about 15–20 minutes of close contact. The incidence of scabies often increases during wars and disasters, such as earthquakes, floods and famines, when people are sleeping and living in very overcrowded situations. Although it is possible to become infected by sleeping in a bed formerly occupied by an infected person, this probably rarely happens. As mites may remain alive on discarded clothing for a few days it is possible that some may be able to infest a new host, but such indirect transmission seems uncommon. Consequently it may not always be necessary to wash or otherwise treat clothing or bedclothes to prevent scabies spreading. However, in epidemics or with cases of *crusted scabies* (Norwegian scabies) clothing and bedding should be dry-cleaned or laundered. Ten minutes at 50 °C will kill the mites.

Increases in the incidence of scabies often appear to go in 15–20-year cycles, probably due to fluctuations in levels of immunity in the human population, although this explanation has been disputed. It is estimated that global prevalence of scabies is at least 300 million cases annually.

18.3   Recognition of scabies

Scabies can be diagnosed by detection of the female mite's narrow twisting tunnels; faeces deposited in these tunnels may be visible through the skin as dark pepper-like spots. Unfortunately tunnels are not always easy to see, especially on dark-skinned people. However, they can often be made visible by smearing a drop of ink on suspected areas of skin infestations; after a few minutes the ink seeps into the tunnels and when the excess is wiped off they become visible. Alternatively if liquid tetracycline is used tunnels fluoresce bright yellow-green under ultraviolet light. Surface layers of skin at the end of the tunnels can be gently scratched away with sharp dissecting needles and the mites, which usually readily adhere to the points of the needles, removed and examined under a ×50 magnification. A cruder procedure is to scrape the affected skin areas with a scalpel blade or razor and to examine the scrapings.

Most people with scabies have only about 11–14 female mites, children 19–20, and only about 3% have more than 50 mites.

18.3.1 The scabies rash

The scabies rash is a papular eruption that occurs mainly on areas of the body not infected with burrowing mites, such as the buttocks and around the waist and shoulders, but the rash can also occur on other parts of the body such as the arms, calves and ankles. It does not appear on the head, centre of the chest or back, nor on the palms of the hands or soles of the feet. The *rash* is in response to an allergic reaction produced by the mites. Frequently patients are unaware they have mites until a rash appears.

When a person is infected for the *first time* with mites the rash does not usually appear until 4–6 weeks, although in exceptional cases the rash may occur within 2 weeks. However, in those who have previously been infected a rash may develop within 2–4 days after reinfection. Severe *pruritus* soon develops and causes vigorous and constant scratching, especially at night and after hot baths. Scratching frequently causes secondary bacterial infections, which may be quite severe resulting in boils, pustules, ecthyma, eczema and impetigo contagiosa. Such complications may hinder the detection of mites, and consequently correct diagnosis of scabies may not be made. The seriousness of the symptoms is not always directly related to the number of mites, and severe reactions may be found on people harbouring few mites. The rash may persist for 2–3 weeks after all scabies mites have been destroyed.

A condition known as *crusted* or Norwegian scabies is highly contagious and although relatively rare seems to be increasing. It is caused by vast numbers, sometimes many thousands, of mites, and up to 4700 mites per gram of skin have been recorded! With such a large mite population the skin becomes scaly and hyperkeratotic, and crusts form over the hands and feet and scaling eruptions on other parts of the body. There is, however, much less pronounced itching. It seems the condition arises due to a loss of immunity, which allows the establishment of enormous numbers of mites. This hypothesis is supported by the development of crusted scabies commonly in patients using corticosteroids, in patients with HIV and in those mentally impaired. In such people irritation is reduced and consequently there is less scratching, an act that normally helps in the removal and destruction of mites.

18.4   Treatment of scabies

In treating scabies a 1% lindane (HCH) cream (Kwell) is very effective in killing the mites, but it should not be used on infants, small children, pregnant or nursing mothers because of its toxicity, and it has been banned in several countries. Equally efficient, and considered by many to be the most effective treatment, is 5% permethrin cream or lotion (Elimite). This should be applied to the skin from the neck downwards and washed off after 8–14 hours. Sometimes a second application a week later is recommended. Crotamiton (Eurax) applied as a 10% cream or lotion is also a very safe treatment, but two to five daily applications are needed. Both crotamiton cream and 10% sulphur ointments may be used on infants and young children. Ivermectin given as a single oral dose of 200 mg/kg body weight is effective in killing scabies mites, and especially useful in treating crusted scabies infections, but in many countries it has not been cleared for such treatments. Resistance to both permethrin and ivermectin has been documented.

Benzyl benzoate has been used for many years, and may still be appropriate in situations when the above acaricidal creams and lotions are too costly. A 20–25% benzyl benzoate emulsion can be painted on a patient from the neck downwards, and after allowing some 5–10 minutes for this application to dry the patient can re-dress. Sometimes the cure rate is only about 50%, so a repeat treatment on the third day may be advisable.

Scabies is highly contagious, so it is important to treat all family members or communities living in close association, such as in nursing homes, not just individuals diagnosed with mites, otherwise reinfestations will occur. In such situations, or when there is crusted scabies, it is advisable to launder or dry-clean clothes and bedding to kill any mites on these items.

Killing the mites will do little to immediately alleviate the irritation caused by the rash, and consequently patients will continue to complain of itchiness until the itching eventually disappears. It may therefore be advisable to prescribe antihistamines to relieve any itching.

Further reading

Arlian, L. G. (1989) Biology, host relations, and epidemiology of *Sarcoptes scabiei*. *Annual Review of Entomology*, **34**, 139–61.

Buffet, M. and Dupin, N. (2003) Current treatments for scabies. *Fundamental and Clinical Pharmacology*, **17**, 217–25.

Burkhart, C. G., Burkhart, C. N. and Burkhart, K. M. (2000) An epidemiologic and therapeutic reassessment of scabies. *Cutis*, **65**, 233–40.

Cox, N. H. (2000) Permethrin treatment in scabies infestation: importance of the correct formulation. *BMJ*, **300**, 37–8.

Daisley, H., Charles, W. and Suite, M. (1993) Crusted (Norwegian) scabies as a pre-diagnostic indicator for HTLV-1 infection. *Transactions of the Royal Society of Tropical Medicine and Hygiene*, **87**, 295.

Glaziou, P., Cartel, J. L., Alizieu, P., Briot, C., Moula-Pelat, J. P. and Martin, P. M. V. (1993) Comparison of ivermectin and benzyl benzoate for treatment of scabies. *Tropical Medicine and Parasitology*, **44**, 331–3.

Marliere, V., Roul, S., Labreze, C. and Taieb, A. (1999) Crusted (Norwegian) scabies induced by use of topical corticosteroids and treated successfully with ivermectin. *Journal of Pediatrics*, **135**, 122–4.

Meinking, T. L. Burkhart, C. N. and Burkhart, C. G. (1999) Ectoparasitic diseases in dermatology: reassessment of scabies and pediculosis. *Advances in Dermatology*, **15**, 77–108.

Meinking, T. L. and Elgart, G. W. (2000) Scabies therapy for the millennium. *Pediatric Dermatology*, **17**, 154–6.

Mellanby, K. (1943) *Scabies*. Middlesex: E. W. Classey. (Reprinted 1972 with minor additions.)

Mullen, G. and O'Connor, B. M. (2002) Mites (*Acari*). In *Medical and Veterinary Entomology*, ed. G. Mullen and L. Durden. Amsterdam: Academic Press, pp. 449–516.

Orkin, M. and Maibach, H. T. (eds.) (1985) *Cutaneous Infestations and Insect Bites*. New York, NY: Marcel Dekker. See Chapters 1–18 on scabies.

Paller, A. S. (1993) Scabies in infants and small children. *Seminars in Dermatology*, **12**, 3–8.

Turner, S., Lines, S., Chen, Y., Hussey, L. and Aguis, R. (2005) Work-related infectious disease reported to the Occupational Disease Intelligence Network and the Health and Occupation Reporting Network in the UK (2000–2003). *Occupational Medicine (London)*, **55**, 275–81.

Walker, G. J. A. and Johnstone, P. W. (2000) Interventions for treating scabies. *Archives of Dermatology*, **136**, 387–9.

19

Scrub typhus mites (Trombiculidae)

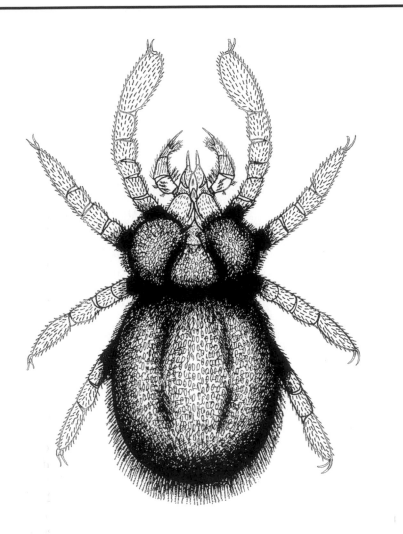

There are thousands of species of trombiculid mites in many genera but only about 20 species commonly attack people. The family Trombiculidae has a more or less worldwide distribution, but the medically most important species, such as *Leptotrombidium deliense*, *L. akamushi* and *L. fletcheri*, which are vectors of scrub typhus *Orientia* (= *Ricksettsia*) *tsutsugamushi*, are found in Asia, the Pacific regions and the north-east coast of Australia.

Other trombiculid mites in many parts of the world cause itching and a form of dermatitis known as scrub itch, autumnal itch or trombidiosis. In northern Europe larvae of *Neotrombicula autumnalis* (**harvest mites**) and in North America and parts of Central and South America larvae of *Eutrombicula alfreddugesi* (**red bugs**) commonly attack people and cause considerable discomfort.

19.1 External morphology

19.1.1 Adults and nymphs

Adults are small (1–2 mm), usually reddish and covered dorsally and ventrally with numerous *feathered hairs* giving them a velvety appearance. The four pairs of legs end in paired claws. The body is distinctly constricted between the third and fourth pairs of legs, giving it an outline resembling a *figure of eight*. Palps and mouthparts project in front of the body and are clearly visible (Fig. 19.1).

Nymphs resemble the adults but are smaller (0.5–1.0 mm) and the body is less densely covered with feathered hairs.

Neither adults nor nymphs are of direct medical importance; they do not bite humans or animals but feed on small arthropods and their eggs. Only the larvae are parasitic and hence disease vectors.

19.1.2 Larvae

Larvae are very small (0.15–0.3 mm), but after engorging may increase six-fold in size. They are usually reddish or orange but may be pale yellowish. There are three pairs of legs, with each leg terminating in a pair of relatively large claws. Both legs and body are covered with fine *feathered hairs*. The five-segmented *palps* and *mouthparts* are large and conspicuous, giving the larva the appearance of having a false head (Fig. 19.2, Plate 24). Dorsally on the anterior part of the body there is a rectangular or pentagonal-shaped *scutum*, but as it is weakly sclerotized it is often difficult to see under the microscope, unless the light is correctly aligned. More easily detected is a pair of eyes on either side of the scutum. In medically important species there are *five feathered setae* on the scutum and in addition a pair of specialized feathered hairs known as *sensillae* which arise from distinct bases. The combination of a body covered with feathered hairs, five scutal hairs, a pair of flagelliform sensillae and large pigmented eyes distinguishes larvae of *Leptotrombidium* from larvae in other mite genera.

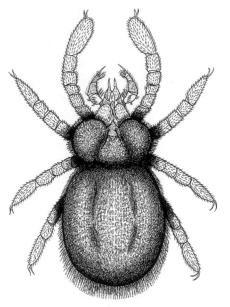

Figure 19.1 Dorsal view of an adult scrub typhus mite (*Leptotrombidium* species).

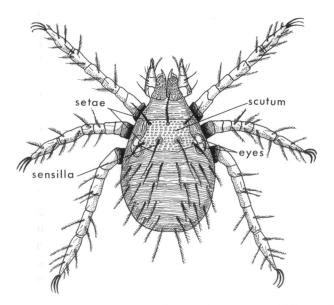

Figure 19.2 Dorsal view of a larval *Leptotrombidium* mite. Note that the scutum has a pair of sensillae and five scutal setae.

19.2   Life cycle

The life cycle of trombiculid mites is complex, and unfortunately several different names have been used for the various larval and nymphal stages (see summarized life cycle on p. 250).

Adult trombiculid mites are not parasitic but live in the soil, feeding on a variety of small soil-inhabiting arthropods and their eggs. Females lay 1–5 spherical eggs each day on leaf-litter and the surface of damp but well-drained soil, such as river banks, scrub-jungle, grassy fields and neglected gardens. In hot climates egg-laying continues uninterrupted for a year or more, but in cooler areas of South-east Asia, including Japan, oviposition ceases during the cooler months and adults enter into partial or complete hibernation.

After about 4–7 days the eggshell splits, but the six-legged larva does not emerge – it remains within the eggshell and is called the *deutovum*. After about 5–7 days the larva crawls out of the eggshell and becomes very active, swarming over the ground and climbing up grasses and other low-lying vegetation. Larvae attach themselves to birds and mammals (especially rodents), and also to people walking through infested vegetation. When on a suitable animal host larvae congregate where the skin is soft and moist, such as the ears, genitalia and around the anus. On people larvae seek out areas where clothing is tight against the skin, such as around the waist or ankles.

Larvae pierce the host's skin and inject saliva into the wound, which causes disintegration of the cells. Larvae do not suck up blood, but feed on lymph and other fluid and semi-digested tissues. Repeated injection of saliva into the wound produces a skin reaction in the host result-ing in the formation of a tube-like structure extending vertically down-wards in the host's skin, which is known as the *stylostome* or *his-tiosiphon*. Some trombiculid mites remain attached to their hosts for up to a month, but the *Leptotrombidium* vectors of scrub typhus remain on people for only about 3–8 days. The engorged larva drops to the ground and buries itself just below the surface of the soil or underneath debris (Fig. 19.3).

Having concealed itself the larva becomes inactive, and this stage is known as the *protonymph*. After 7–10 days the protonymph moults to pro-duce an eight-legged reddish *deutonymph* covered with feathered hairs. These nymphs are not parasitic, but feed on soil-inhabiting arthropods. After a few days to about two weeks the deutonymph ceases feeding, becomes inactive and is called a *tritonymph*, which after about another 14 days moults to give rise to an adult. Adults resemble nymphs but are larger, and like them are free-living and feed on small soil-inhabiting animals.

The life cycle (Fig. 19.3) usually takes 40–75 days but may be as long as 8–10 months. The stages in the life cycle can be summarized as

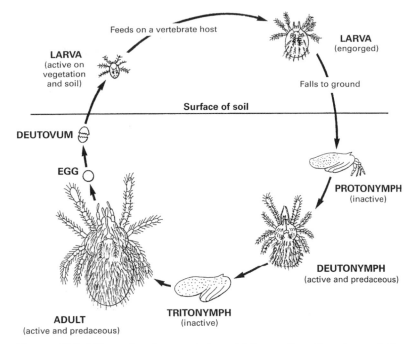

Figure 19.3 Life cycle of a *Leptotrombidium* mite. (Courtesy of the Natural History Museum, London.)

follows (the inactive stages are bracketed, and alternative names are in italics):

(Egg) → (Deutovum, *prelarva*) → Larva → (Protonymph, *first nymphal stage, nymphochrysalis*) → Deutonymph, *second nymphal stage* → (Tritonymph, *third nymphal stage, preadult, imagochrysalis*) → Adult

19.3   Ecology

Free-living nymphs and adults need habitats containing sufficient numbers of suitable arthropods to serve as food, while the larvae require habitats to have small mammals so that they will have hosts on which to feed. **Rodents** such as species of *Rattus*, *Apodemus* and *Microtus* and insectivores such as shrews (*Suncus* species) and tree shrews (*Tupaia* species) are important larval hosts. It seems that domestic rats play little or no part in the ecology of scrub typhus. In addition to these hosts other more or less incidental hosts, including birds, may be important in aiding the dispersal of larvae to other areas.

Areas supporting large populations of mites are frequently formed when people clear away vegetation for agricultural purposes or for the collection of firewood, that is activities which create **scrub** vegetation, from which both vector and infection derive their names. Such habitats favouring colonization by *Leptotrombidium* mites are called '**mite islands**'. These may be very small, often just a few square metres, or they may comprise several square kilometres. This can result in very patchy and isolated distributions of *Leptotrombidium* mites and consequently areas where people are exposed to infection.

19.4   Medical importance

19.4.1 Nuisance
Several species of trombiculid mites attack people in temperate and tropical regions. In northern Europe the main pest is the harvest mite (*Neotrombicula autumnalis*), while in the USA it is the red bug (*Eutrombicula alfreddugesi*). Although these mites do not transmit infections they can nevertheless cause intense itching and irritation, commonly referred to as 'harvest-bug itch', 'autumnal itch' or 'scrub itch'. Larval mites commonly attack the legs. If they are forcibly removed, their mouthparts frequently remain embedded in the skin and this may promote further irritation. People usually become infested with these mites after walking through long grass or scrub vegetation, especially in the autumn or summer.

19.4.2 Scrub typhus
The causative organism of scrub typhus is the rickettsia *Orientia tsutsugamushi*, and the disease is known as scrub typhus, mite-borne typhus, Japanese river fever, chigger-borne rickettsiosis or tsutsugamushi disease. The disease is restricted to the Asia–Pacific area, extending from the Primorye region of Siberia through Pakistan and India to Myanmar, Indonesia, Malaysia, Thailand, South-east Asia, China, Taiwan, the Philippines, Japan, Papua New Guinea, New Zealand, north-eastern Australia and neighbouring south-west Pacific islands. Although scrub typhus is mostly reported from low-lying areas, it can occur at 1000 m in many areas, and has been reported up to about 2000 m in Taiwan and 3200 m in the Himalayas.

During World War II (1939–45) the incidence of scrub typhus in troops in the Asia–Pacific area was second only to malaria. Several species of mites are known or suspected as being vectors, but only about seven are important. In most areas the principal vectors are *Leptotrombium deliense* and *L. akamushi*, but several other species can be important local vectors, such as *L. fletcheri* in Malaysia.

People become infected by the bites of larval trombiculid mites when they visit or work in areas having so-called mite islands, that is patches of vegetation harbouring large numbers of host-seeking larvae. The disease is often associated with 'fringe habitats', in other words habitats separating two major vegetation zones such as forests and plantations, because such areas are often heavily populated with rodent hosts. Consequently the risk of scrub typhus transmission is often associated with areas having different types of vegetation, that is habitat diversity.

Because larvae attach themselves to only a *single host* during their life-time scrub typhus cannot be spread by larvae feeding on one infected host (e.g. humans) and then another. Infection acquired by larval mites feeding on hosts with *O. tsutsugamushi* is passed on to the free-living nymphal stages and then to the free-living adults. When a female lays eggs they are infected with rickettsiae and this infection is passed to the emerging larvae. So, although larvae have not previously fed on humans they are already infected and consequently transmit the disease to their hosts (humans or rodents) when they feed for the first and only time. In addition to such *transovarial* transmission there is also *co-feeding* transmission. That is, when a larva is feeding very near an infected larva which is also feeding, the rickettsiae being injected into the host from the infected larva can be picked up by the previously uninfected larva. This acquired infection can then be passed through subsequent immature stages to adult mites, females of which will lay rickettsia-infected eggs.

Leptotrombidium mites themselves are the main *reservoirs* of infection.

19.5   Control

Repellents applied to the skin, such as DEET, dimethyl carbamate or benzyl benzoate, can help reduce the likelihood of people becoming infested with mites. Clothing, especially socks and trousers, can also be impregnated or sprayed with repellents or permethrin.

The patchy distribution of chigger mites makes their control very difficult. However, if mite islands are identified then scrub vegetation can sometimes be destroyed mechanically, with herbicides or by burning. This approach is not practical if mites are inhabiting cultivated land where ground vegetation consists mainly of crops. When insecticides (acaricides) such as carbaryl, propoxur, permethrin or deltamethrin are sprayed on vegetation harbouring mites they can be substantially reduced in number. Such applications can be from ground-based sprayers or from aircraft. Ultra-low-volume (ULV) spraying can also be undertaken. However, in some situations such blanket insecticidal coverage of vegetation may not be environmentally acceptable.

Further reading

Azad, A. F. (1986) Mites of public health importance and their control. Mimeographed document WHO/VBC/86.931. Geneva: World Health Organization.

Frances, S. P. and Khlaimanee, N. (1996) Laboratory tests of arthropod repellents against *Leptotrombidium deliense* – noninfected and infected with *Rickettsia tsutsugamushi* – and noninfected *L. fletcheri* (Acari: Trombiculidae). *Journal of Medical Entomology*, **33**, 232–5.

Frances, S. P., Watcharapichat, D., Phulsuksombati, D. and Tanskul, P. (2000) Transmission of *Orientia tsutsugamushi*, the aetiological agent for scrub typhus, to co-feeding mites. *Parasitology*, **120**, 601–7.

Hengbin, G., Min, C., Kaihua, T. and Jiaqi, T. (2006) The foci of scrub typhus and strategies of prevention in the spring in Pingtan Island, Fujian Province. *Annals of the New York Academy of Sciences*, **1078**. 188–96.

Kawamura, A., Tanaka, H. and Tamura A. (eds.) (1996) *Tsutsugamushi Disease: an Overview*. Tokyo: University of Tokyo Press.

Mount, G. A., Grothaus, R. H., Baldwin, K. F. and Haskings, J. R. (1975) ULV sprays of propoxur for control of *Trombicula alfreddugesi*. *Journal of Economic Entomology*, **68**, 761–2.

Ogawa, M., Hagiwara, T., Kishimoto, T. *et al.* (2002) Scrub typhus in Japan: epidemiology and clinical features of cases reported in 1998. *American Journal of Tropical Medicine and Hygiene*, **67**, 162–5.

Roberts, S. H. and Zimmerman, J. H. (1980) Chigger mites: efficacy of control with two pyrethroids. *Journal of Economic Entomology*, **73**, 811–2.

Sasa, M. (1961) Biology of chiggers. *Annual Review of Entomology*, **6**, 221–44.

Strickman, D. (2001) Scrub typhus. In *The Encyclopedia of Arthropod-Transmitted Infections of Man and Domesticated Animals*, ed. M. W. Service. Wallingford: CABI, pp. 456–62.

Takahashi, M., Misumi, H., Urakami, H. *et al.* (2004) Mite vectors (Acari: Trombiculidae) of scrub typhus in the new endemic area in northern Kyoto, Japan. *Journal of Medical Entomology*, **41**, 107–14.

Takahashi, M., Murata, M., Misumi, H., Hori, E., Kawamura, A. and Tanaka, H. (1994) Failed vertical transmission of *Rickettsia tsutsugamushi* (Rickettsiales: Rickettsiaceae) acquired from rickettsemic mice by *Leptotrombidium pallidum* (Acari: Trombiculidae). *Journal of Medical Entomology*, **31**, 212–16.

Traub, R. and Wisseman, C. L. (1968) Ecological considerations in scrub typhus. 1. Emerging concepts. *Bulletin of the World Health Organization*, **39**, 209–18.

Traub, R. and Wisseman, C. L. (1968) Ecological considerations in scrub typhus. 2. Vector species. *Bulletin of the World Health Organization*, **39**, 219–30.

Traub, R. and Wisseman, C. L. (1968) Ecological considerations in scrub typhus. 3. Methods of area control. *Bulletin of the World Health Organization*, **39**, 231–7.

Traub, R. and Wisseman, C. L. (1974) The ecology of chigger-borne ricksettsiosis (scrub-typhus). *Journal of Medical Entomology*, **11**, 237–303.

Walter, D. E. and Proctor, H. C. (1999) *Mites: Ecology, Evolution and Behaviour*. Sydney: University of New South Wales Press.

20

Miscellaneous mites

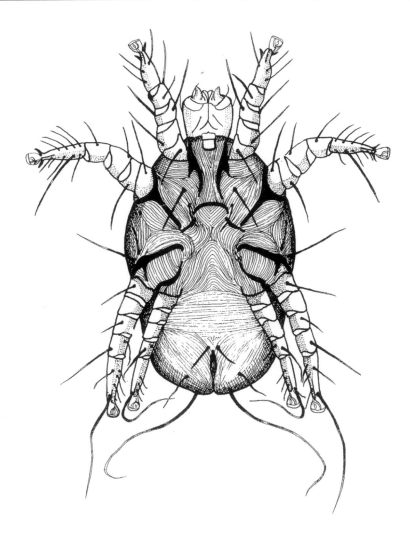

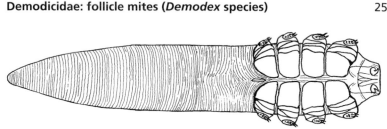

Figure 20.1 Ventral view of hair follicle mite (*Demodex folliculorum*), showing four pairs of stumpy legs.

In addition to scabies mites (Chapter 18) and scrub typhus mites (Chapter 19) there are many other species of mites that can be of medical importance. The most important two, the follicle mites and the house-dust mites, are described below, followed by very brief mentions of a few other mites.

20.1   Demodicidae: follicle mites (*Demodex* species)

Two species of *Demodex* commonly infect humans, namely *Demodex folliculorum* (Fig. 20.1) and *D. brevis*. The former is the more elongate species (0.2–0.4 mm long) and primarily inhabits hair follicles and eyelash hair follicles, whereas *D. brevis* is squatter (0.15–0.2 mm) and lives in the sebaceous glands of hairs and eyelashes. A single follicle may contain 25 *D. folliculorum*, but sebaceous glands contain many fewer *D. brevis*. Both species have a striated body and four pairs of very short stubby legs; they are remarkably non-mite-like.

Demodex mites feed on subcutaneous tissues, especially sebum, and are particularly common on the forehead, nose, eyelids and cheeks adjacent to the nose. Eggs hatch to produce six-legged *larvae* which moult to give rise to *protonymphs* then nymphs and finally adults. All the developmental stages, which extend over 13–15 days, occur within the hair follicles or sebaceous glands. Transfer of mites is believed to occur between mother and infants during the close contact of nursing. Incidence of infection increases with age, and it seems that 90–100% of older people have these mites!

However, most people are unaware they have mites because they have not produced any adverse effects, and control methods are therefore unnecessary. But occasionally mites cause eruptions on the face such as acne, rosacea, impetigo contagiosa or blepharitis. Repeated daily washing with a baby shampoo diluted to 50% followed by antibiotic cream may reduce infections, especially around the eyelashes. For more serious cases, 5% selenium sulphide cream, 10% sulphur or 1% mercury oxide ointment have been used, although often with poor results. More recently 5% permethrin cream or 10% sodium sulfacetamide formulations have given better results. In trials oral dosages of ivermectin have shown promise in killing the mites.

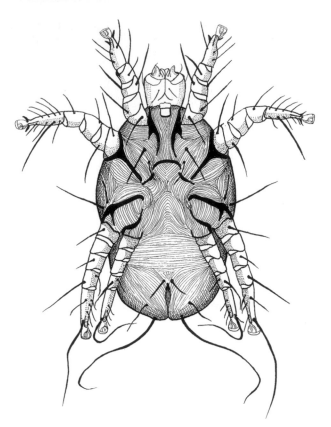

Figure 20.2 Ventral view of a house-dust mite (*Dermatophagoides pteronyssinus*).

20.2   Pyroglyphidae: house-dust mites (*Dermatophagoides* and *Euroglyphus* species)

More than 20 species of mites are found in house dust. The most common are *Dermatophagoides pteronyssinus*, known as the European house-dust mite, and *D. farinae*, the American house-dust mite, but both species are found more or less worldwide. Another house-dust mite, *Euroglyphus maynei*, also has an almost worldwide distribution.

Dermatophagoides (Fig. 20.2) and *Euroglyphus* mites are very small (0.3 mm) and live among bedclothes, mattresses, carpets and general house dust. Female mites lay about 1–3 eggs a day. These hatch after 6–12 days and a six-legged *larva* emerges, which feeds and passes through two *nymphal* stages before becoming an adult. The complete life cycle takes about 3–4 weeks. *Beds* are the most important, and sometimes only, breeding site. Mites feed on fungi growing on floors and mattresses, discarded skin scales, semen and other organic debris. *High humidities* are required for mite survival and breeding. Adults live about 1–2 months and produce 40–80 eggs. They are very rarely seen but become airborne when beds are made and are easily inhaled, together with their faeces, and this can cause allergic symptoms resulting in asthma, rhinitis and eczema.

Typically there may be 300 mites per gram of house dust. Under ideal breeding conditions some 5000 mites may be found in just one gram of mattress dust, but up to 15 600 has been recorded! *Densities* above 100 mites per gram are considered a risk factor for sensitization to allergies such as rhinitis, conjunctivitis, atopic eczema and asthma, while 500 mites per gram is a major risk factor in the development of acute asthma in those allergic to house-dust mites.

Vacuum-cleaning carpets may have little effect on removing live mites because they cling firmly to carpet fibres, although it may be better in removing mites from mattresses and sheets. However, vacuum-cleaning may increase allergies by creating mite-infested dust storms. Enclosing mattresses and pillows in plastic covers can help reduce mite infestations, although sleeping under such conditions may be uncomfortable. Washing bedding in water above 55 °C or dry-cleaning will kill mites, as will leaving electric blankets switched on at maximum temperature for six hours or more during the day. Treating sheets with 5% benzyl benzoate can be effective in killing mites, as can spraying carpets with pirimiphos-methyl or permethrin. Insecticide-treated mattress covers have been promoted, but such impregnated items may themselves cause allergies in sensitive people, including those already suffering from mite allergies.

It can prove very difficult to reduce mite densities sufficiently to give any reduction in allergies.

20.3 Other mites

Numerous other mites are parasitic on mammals and birds, including pets and livestock, and also insects, and some occasionally become parasitic on people. For example, *Pyemotes tritici*, known as the grain mite or hay or straw itch mite, usually parasitizes larvae of grain moths and beetles, but can bite people handling infested grain and straw. The cosmotropical rat mite (*Ornithonyssus bacoti*), the tropical fowl mite (*O. bursa*) and the chicken mite (*Dermanyssus gallinae*) may sometimes attack people, especially those working closely with infested animals. Bites from these mites can cause irritation and dermatitis. A few species such as *Liponyssoides sanguineus*, which normally is ectoparasitic on mice and rats, can transmit rickettsial pox (*Rickettsia akari*), and this mite and a few other species can also transmit Q fever (*Coxiella burnetii*) to people.

There are also other mites, such as forage mites that live among stored products such as grain, copra, flour, dried fruits, cheese and animal feeds.

People habitually handling these substances may develop allergic symptoms such as dermatitis, and more rarely bronchitis and asthma. This may lead to terms such as 'grocer's itch', 'baker's itch', and 'copra itch' to describe these occupational hazards.

Further reading

Akilov, O. E. and Mumcuoglu, K. Y. (2004) Immune response in demodicosis. *Journal of the European Academy of Dermatology and Venereology*, **18**, 440–4.

Boner, A., Pescollderungg, L. and Silverman, M. (2002) The role of house dust mite elimination in the management of childhood asthma: an unresolved issue. *Allergy*, **57** (Suppl. 74), 23–31.

Cameron, M. M. (1997) Can house dust mite-triggered atopic dermatitis be alleviated using acaricides? *British Journal of Dermatology*, **137**, 1–8.

Cloosterman, S. G. M. and van Schayck, O. C. P. (1999) Control of house dust mite in managing asthma: effectiveness of measures depends on stage of asthma. *BMJ*, **318**, 870.

Colloff, M. J., Ayres, J., Carswell, F. *et al.* (1992) The control of allergens of dust mites and domestic pets: a position paper. *Clinical and Experimental Allergy*, **22**, 1–28.

Fain, A., Guérin, B. and Hart, B. J. (1990) *Mites and Allergic Disease*. Varennes en Argonne, France: Allerbio.

Harvey, P. and May, R. (1990) Matrimony, mattresses and mites. *New Scientist*, 3 March, 48–9.

Herron, M. D., O'Reilly, M. A. and Vanderhooft, S. L. (2005) Refractory *Demodex* folliculitis in five children with acute lymphoblastic leukemia. *Pediatric Dermatology*, **22**, 407–11.

Krantz, G. W. (1978) *A Manual of Acarology*, 2nd edn. Corvallis, OR: Oregon State University.

Løvik, M., Gaarder, P. I. and Mehl, R. (eds.) (1998) The house-dust mite: its biology and role in allergy. Proceedings of an international scientific workshop, Oslo, Norway, 4–7 September 1997. *Allergy* **53** (Suppl. 48), 1–135.

McDaniel, B. (1979) *How to Know the Mites and Ticks*. Dubuque, IA: W. C. Brown.

Mumcuoglu, Y. (1976) House dust mites in Switzerland: I. Distribution and taxonomy. *Journal of Medical Entomology*, **13**, 361–73.

Nutting, W. B. (ed.) (1984) *Mammalian Diseases and Arachnids. Volume 1: Pathogen Biology and Clinical Management. Volume 2: Medico-Veterinary, Laboratory, and Wildlife Diseases, and Control.* Boca Raton, FL: CRC Press.

Owen, S., Morganstern, M., Hepworth, J. and Woodcock, A. (1990) Control of house dust mite antigen in bedding. *Lancet*, **335**, 396–7.

Rosen, S., Yeruham, I. and Braverman, Y. (2002) Dermatitis in humans associated with the mites *Pyemotes tritici, Dermanyssus gallinae, Ornithonyssus bacoti* and *Androlaelaps casalis* in Israel. *Medical and Veterinary Entomology*, **16**, 442–4.

Siebers, R., Nam, H.-S. and Crane, J. (2004) Permeability of synthetic and feather pillows to live house dust mites and house dust. *Clinical and Experimental Allergy*, **34**, 888–90.

Walter, D. E. and Shaw, M. (2005) Mites and disease. In *Biology of Disease Vectors*, 2nd edn, ed. W. C. Marquart. Amsterdam: Elsevier Academic Press, pp. 25–44.

Wharton, G. W. (1976) House dust mites. *Journal of Medical Entomology*, **12**, 577–621.

Appendix   Names of some chemicals and microbials used in vector control (with common trade names in parentheses)

Carbamates
Bendiocarb (Ficam)
Carbaryl (Sevin)
Propoxur (Baygon, Arprocarb)

Formamidines
Amitraz (Mitaban)
Chlordimeform

Inorganics
Borax
Paris Green

Insect growth regulators (IGRs)
Cyromazine
Diflubenzuron (Dimilin)
Fenoxycarb
Hydrophen
Hydroprene
Kinoprene
Lufenuron
Methoprene (Altosid)
Novaluron (Rimon)
Noviflumuron
Pyriproxyfen
Triflumuron

Microbials
Bacillus sphaericus (Vectolex)
Bacillus thuringiensis var. *israelensis* (*Bti*) (= *Bacillus thuringiensis* H-14)
(Aquabac, Teknar, Vectobac)
Spinosad (fermentation of an actinomycete bacterium, *Saccharopolyspora spinosa*)

Monomolecular films
Isostearyl alcohols (Arosurf MSF, Agnique MMF)
Lecithins

Natural organics
Neem oil
Petroleum oils
Pyrethrum

Nicotinoids
Dinotefuran
Imidacloprid

Organochlorines
DDT
Endosulfan
Gamma-HCH (Lindane)
HCH (formerly BHC)
Methoxychlor
Toxaphene (Camphechlor)

Organophosphates
Bromophos
Chlorpyrifos (Dursban)
Coumaphos
Diazinon
Dimethoate
Dioxathion (Hercules)
Fenchlorphos (Fenchlorvos, Ronnel)
Fenitrothion (Sumithion, Nuvanol)
Fenthion (Baytex)
Iodofenphos (Iodfenphos, Jodfenphos)
Malathion
Naled (Dibrom)
Phoxim
Pirimiphos-methyl (Actellic)
Temephos (Abate)
Trichlorphon (Dipterex)

Pheromones
Muscalure

Phenylpyrazoles
Fiprinol (Frontline, Top Spot, Termidore)

Pyrethroids
Allethrin
Alpha-cypermethrin
Bioallethrin
Cyfluthrin
Cypermethrin (Cymbush)
Deltamethrin (Decis)
Lambda-cyhalothrin (Icon)
Permethrin (Ambush)
Phenothrin (Sumithrin)
Tetramethrin

Repellents
Autan (Cutter Advanced) – a piperidine-based compound
Bayrepel (KBR or KBR 3023) – a piperidine-based compound
Benzyl benzoate (Ascabiol)
Citronella oil
DEET, i.e. diethyl-methylbenzamide (formerly called diethyltoluamide)
Dibutyl phthalate
Dimethyl carbamate
Lemon eucalyptus oil
PMD – a botanical compound (para-menthane 3,8-diol) derived from lemon eucalyptus

Sulphonamides
Sulfluramid

Synergists
Piperonyl butoxide

Glossary of common terms relevant to medical entomology

This glossary contains terms that are used in this book and a few others that are not, but which are considered pertinent to vector biology and control. In general morphological names are excluded as these are adequately explained in the various chapters. As usual with glossaries the inclusion and exclusion of terms is somewhat subjective.

Acaricide Chemical that kills mites and ticks. Most acaricides are also insecticides.

Aestivation Condition in which an organism has a period of inactivity as an adaptive response to unfavourable conditions encountered during the summer, or in the tropics the hot or dry seasons, e.g. aestivating adults of some *Anopheles* species.

Afrotropical region (formerly Ethiopian zoogeographical region) The countries of Africa south of mid-Sahara, including southern Arabia, Madagascar, Seychelles and Cape Verde Islands. *See* Zoogeographical regions.

Amastigote Morphological form of species of *Leishmania* and *Trypanosoma* with a rounded body and without a flagellum that occurs predominantly in macrophages (*Leishmania* species) or muscle cells (*T. cruzi*) of a vertebrate host.

Amplifying host Usually used in relation to arboviruses where a host attains a very high level of parasites in its blood (high viraemic titre) which makes it very infective to vectors. For example, pigs develop high viraemias of Japanese encephalitis and are important amplifying hosts, as are monkeys in the transmission of Kyasanur Forest disease. Often not an efficient host for long-term maintenance of parasite populations.

Anautogenous Refers to females of blood-sucking insects that require at least one blood-meal to develop their eggs. *See* Autogenous.

Anthropophagic (anthropophilic) Refers to blood-sucking arthropods that prefer to feed on humans rather than other hosts. The degree of anthropophagy can vary geographically within species as well as between species. *See* Host preference.

Aperiodic (non-periodic) Not exhibiting periodicity. When applied to microfilariae of helminths it means they are neither periodic nor subperiodic in their appearance in peripheral vertebrate blood, e.g. *Onchocerca volvulus*.

Arbovirus From *ar*thropod-*b*orne *virus*. A virus that multiplies in a blood-sucking arthropod and is principally transmitted by the bite of arthropods to vertebrate hosts, e.g. yellow fever virus. Viruses, such as myxoma virus in rabbits, transmitted mechanically by arthropods (e.g. fleas and mosquitoes) and involving no cyclical development in the vector are not arboviruses.

Australasian region Australia, Tasmania, New Zealand, Melanesia, Micronesia and Polynesia. *See* Zoogeographical regions.

Autogenous Refers to females of blood-sucking insects (usually mosquitoes and simuliids) that lay at least the first batch of eggs without a blood-meal. Subsequently a blood-meal is required for each further oviposition. Some mosquitoes, e.g. *Toxorhynchites* species, never blood-feed and are thus always autogenous. *See* Anautogenous.

Biodegradable insecticide Insecticide that does not accumulate in the environment but is broken down by micro-organisms in the soil and water into relatively harmless compounds, e.g. the organophosphates and carbamates.

Biological control (biocontrol) Deliberate introduction, or augmentation, of biological agents such as pathogens, parasites and predators (especially fish) to control arthropod populations, mainly mosquito larvae. Although *Bacillus thuringiensis* var. *israelensis* is often included in this category it is not a biological control agent because it does not recycle and persist in the environment but is applied as a non-living formulation and is thus more accurately described as a microbial insecticide.

Biological transmission Transmission of disease organisms with biological involvement between the vector and parasite. Can involve (1) multiplication without change in form, e.g. *Yersinia pestis* in fleas, arboviruses in ticks and mosquitoes; (2) multiplication with change in form, e.g. *Plasmodium* species in *Anopheles*; or (3) no multiplication (usually a decrease in parasite numbers) but with change in form, e.g. filarial parasites in mosquitoes and black-flies.

Borrow pit Excavation, often by the side of a road or railway, dug to provide earth for buildings or embankments. These fill with water and become mosquito larval habitats.

Bromeliad An epiphytic plant growing on trees in the Americas. Its water-filled leaf axils are colonized by mosquito larvae. *See* Epiphyte.

Campestral In epidemiology, used to describe transmission occurring in fields and open spaces, such as plague transmission among wild rodents. But also used sometimes to include transmission in woods and forests, which strictly is sylvatic transmission.

Carbamates Synthetic insecticides which are derivatives of carbamic acid, e.g. carbaryl and propoxur.

Chemosterilant Chemical used to induce sterility, but not usually death, in arthropods so as to control them, e.g. apholate and tepa. Chemosterilized insects are sometimes used in genetic control of insect vectors.

Chitin A constituent of arthropod cuticle giving it a hard or leathery texture. Some insect growth regulators (IGRs) prevent chitin formation and are used in the control of vectors.

Co-feeding Sometimes an infection can be passed from an infected vector feeding on a host to a nearby uninfected vector. This phenomenon has been recorded in mosquitoes, ticks and mites, such as in the transmission of scrub typhus.

Commensal Animals living together or in close association, e.g. roof rats (*Rattus rattus*) living in and around human habitations.

Cytoform (cytotype) A cytologically defined population with a distinctive chromosomal complement, that may be a species or considered as a chromosomal variant within a species. In the latter case individuals are often given vernacular geographical names, e.g. *Anopheles gambiae* Savanna and Forest forms. Different cytoforms may exhibit differences in behaviour.

Dead-end host Host that although infected with an arbovirus, or other pathogen, and maybe severely affected, has a viraemia too low in titre to infect blood-feeding arthropods. Examples are humans in the epidemiology of Japanese encephalitis, and horses in western equine encephalitis transmission.

Definitive host Host in which parasites reach maturity. This rarely occurs in arthropod vectors, but the noted exception is the development of malarial parasites, involving sexual reproduction, in mosquitoes. *See* Intermediate host.

Diapause State of inactivity or arrested development which allows an organism to survive a period of unfavourable conditions. Development cannot be resumed, even under favourable conditions, until diapause is 'broken' by environmental stimuli such as changes in photoperiod (length of daylight) or temperature.

Dichoptic A condition of the head in adult Diptera in which the eyes are widely separated from each other, as, for example, in female adult simuliids. *See* Holoptic.

Diel periodicity Periodicity that occurs about every 24 hours.

Diurnal Refers to activity during the daylight hours, such as blood-feeding in simuliids and appearance in peripheral vertebrate blood of microfilariae of some helminth species, e.g. *Loa loa*.

Emulsion Suspension of minuscule droplets of one liquid in another. For example, when an insecticide dissolved in oil is mixed with water this results in a milky emulsion.

Endemic Describes a disease in a human population that is constantly present and more or less stable. The numbers of new cases cannot exceed the numbers of new susceptible individuals entering the population, but may be fewer. The opposite is epidemic. In animal populations the equivalents are enzootic and epizootic.

Endophagic   Describes insects, such as some mosquitoes, that enter houses to blood-feed.

Endophilic   Describes insects, such as some mosquitoes, that rest in houses before or after blood-feeding in houses or outside.

Endotoxin   Toxin formed inside Gram-negative bacteria that is released only when the bacteria are lysed (broken down). An example is the toxins in the parasporal body of *Bacillus thuringiensis* var. *israelensis*, which on ingestion are lethal to simuliid and mosquito larvae.

Epidemic   Occurrence of a disease in the human population where the numbers of cases exceed the normal expected number of cases. An epidemic situation can be only temporary. *See* Endemic.

Epimastigote (crithidial form)   Morphological form of a trypanosome with the flagellum emerging about halfway in the body but remaining attached to the cell membrane, e.g. trypanosomes in the salivary glands of tsetse-flies and mid-gut of triatomine bugs. This is not the stage that is infective to the vertebrate host. *See* Metacyclic trypanosome.

Epiphyte   Plant that grows on other plants, usually trees, but without being parasitic. It derives its water mainly from rain. Bromeliads, the water-filled leaf axils of which provide mosquito larval habitats, are epiphytes.

Eurasia   A geographical area comprising the continental land mass of Europe and Asia combined.

Exophagic   Term applied to insects that blood-feed outside houses, e.g. *Aedes aegypti* and simuliid species.

Exophilic   Term applied to blood-sucking insects that rest outside houses, irrespective of whether they have fed inside or outside houses.

Exoskeleton   Outer body layer, often called the integument or cuticle, of arthropods that is shed during moulting from one life-stage to another, e.g. third-instar mosquito larva to fourth-instar, and then fourth-instar larva to pupa.

Extrinsic incubation period   Duration of the part of a parasite's life cycle that is completed in a vector, that is the time from a vector becoming infected to it being infective (i.e. capable of transmitting the parasite).

Genetic control   Special type of biological control that uses genetic techniques to control pest populations. Specifically the reduction of the reproductive potential of vectors. For example, the release of sterilized male vectors into field populations which results in producing large numbers of mated, but infertile, females. These females lay eggs that cannot hatch, so causing a population decrease. *See* Sterile male release.

Gonotrophic cycle   Time from first blood-feeding to oviposition, and subsequently between successive ovipositions. The first such gonotrophic cycle may be a day or two longer than subsequent ones. Sometimes the gonotrophic cycle is defined as time from blood-feeding to blood-feeding. This cycle is sometimes called the ovarian cycle.

Habitat Usually means the physical environment in which an animal lives, e.g. the skin in the case of scabies mites, streams for simuliid larvae and animal nests for many ixodid ticks.

Haemocoel Main body cavity of arthropods in which insect blood (haemolymph) circulates.

Haemolymph Insect blood, usually colourless and clear, which fills the haemocoel.

Hemelytron (pl. hemelytra) Fore-wing of certain insects, e.g. triatomine bugs, which has a thickened basal part and a membranous distal part.

Hemimetabolous (incomplete metamorphosis) Describes the development from egg to adult which is gradual, passing through one or more nymphal stages, e.g. lice and ticks. If adults are winged the nymphal wing buds grow larger at each moult, e.g. bedbugs. There is no pupal stage. *See* Holometabolous.

Hereditary transmission Involves a female vector passing disease organisms to her eggs and thus to the next generation, i.e. transovarial transmission.

Hibernation Period of inactivity and/or altered behaviour caused by cold conditions, e.g. winter. Some mosquitoes, e.g. *Culex pipiens* in Europe, may remain in complete hibernation without blood-feeding for many months by living off their fat reserves. Other species, e.g. *Anopheles atroparvus*, enter incomplete hibernation, during which time adults need to emerge periodically from hibernation sites to take blood-meals to renew their fat reserves.

Holarctic Sometimes used to encompass the Palaearctic and Nearctic regions. *See* Zoogeographical regions.

Holometabolous (complete metamorphosis) Arthropod development from egg to adult in which the body form changes completely in appearance. A pupal or puparial stage is characteristic of holometabolous development. *See* Hemimetabolous.

Holoptic A condition of the head of adult Diptera in which the eyes meet or nearly meet each other, as, for example, in male adult simuliids. *See* Dichoptic.

Host preference Preferred hosts (e.g. species, sex, age) of an arthropod in an area when a choice exists. For a particular species this can alter from area to area as well as seasonally, depending usually on the availability of alternative hosts.

IGRs (insect growth regulators) Sometimes known as insect development inhibitors, these are groups of chemicals that either prevent the development of larvae into pupae or pupae into adults (juvenile hormone analogues, e.g. methoprene) or interfere with the moulting process, killing larvae as they moult (chitin synthesis inhibitors, e.g. diflubenzuron).

Impoundment Well-defined man-made area of water, usually with more or less vertical sides, often constructed to provide water for domestic, agricultural or recreational puposes. However, impoundments are dug sometimes principally to reduce mosquito breeding in formerly marshy areas.

Infected   Applied to arthropods when a parasitic infection has been taken up by the vector but is not yet in a stage in which it can be transmitted to a host. Examples are simuliid adults with onchocercal larvae in their flight muscles, or a vector with arboviruses in the stomach but not yet in the salivary glands. *See* Infective.

Infective   Applied to arthropods when the parasites can be transmitted by the vector to a host. Examples are mosquitoes with malarial sporozoites in the salivary glands or simuliids with third-stage onchocercal worms in their mouthparts. *See* Infected.

Insecticide resistance   Ability of arthropods to tolerate doses of insecticide which would prove lethal to the majority of normal (susceptible) individuals of the same species. Rare mutants which are resistant are selected for by the use of the insecticide.

Instar   One of a series of life-cycle stages in metamorphosis that are separated by a moult, e.g. the first, second and third larval instars of houseflies and the five nymphal stages of bedbugs.

Integument   The cellular epidermis and outer non-cellular cuticle which together provide the outer covering of arthropods. *See* Exoskeleton.

Intermediate host   Host in which a parasite does not reach sexual maturity. Applies to most parasites in arthropod vectors, e.g. filarial parasites in mosquitoes and simuliids. *See* Definitive host.

Intrinsic incubation period   Duration of the life cycle of a parasite in the vertebrate host; interval between infection and first clinical symptoms.

Karyotype   The number and appearance of the chromosomes in the nuclei of a species.

Larviparous   Reproduction in which the egg(s) hatch within the female and larva(e) are deposited, e.g. tsetse-flies.

Life cycle (life history)   In entomology and parasitology this usually means the series of morphological stages an organism passes through to reach the mature adult stage, and the biology of each stage.

Longevity   How long an organism lives, often expressed as the mean expectancy of life. Vector longevity is one of the most important factors in disease transmission dynamics and vector control.

Maggot   Legless larva that has no distinct head, thorax or abdomen, e.g. larvae of house-flies.

Maintenance host   A vertebrate or arthropod host which allows long-term survival of parasite populations. The host must have an infection rate that is at least adequate to maintain a population of the disease agent continuously endemic in an area. Humans can be maintenance hosts of louse-borne typhus, and mosquitoes and birds appear to be maintenance hosts of some of the encephalitis viruses. A maintenance host may retain an infection during periods when there is no or little vector transmission, e.g. during dry seasons or winters.

Mechanical transmission Transmission where there is no multiplication or cyclical development of the aetiological agent (i.e. parasite or pathogen), it being merely passively carried by the vector. Examples are house-flies transmitting trachoma virus and dysenteries by their feet, vomit or faeces, and trypanosomes transmitted by stable-flies (*Stomoxys* species).

Metacyclic trypanosome (metatrypanosome) The final, and usually smaller, version of the trypomastigote form in the vector that is infective for the vertebrate host.

Metamorphosis Changes in form from the first stage (egg) in the life cycle of an arthropod to the adult form. In hemimetabolous arthropods, e.g. bedbugs, lice, ticks and mites, the change is gradual through nymphal stages which resemble small versions of the adult. In holometabolous arthropods, e.g. mosquitoes, tsetse-flies and fleas, the changes are abrupt and involve larval stage(s) and a pupal or puparial stage which are very dissimilar to the adult.

Microbial insecticide Insecticide comprising a biological agent such as bacteria, e.g. *Bacillus thuringiensis* var. *israelensis* and *B. sphaericus*, or toxic compounds derived from such agents.

Middle East Many definitions, but here covers land surrounding southern and eastern shores of the Mediterranean Sea, extending from Morocco to the Arabian Peninsula, including Egypt, Turkey, Iran, Iraq, Sudan and Libya.

Monolayer Thin film of a water-insoluble surfactant, e.g. isostearyl alcohols and lecithins, that have a very high spreading power on water. Mosquito larvae are unable to maintain normal contact with the water surface, and this combined with the reduction of dissolved oxygen content caused by monolayers causes their death.

Morphology The outward structure of an organism. Most arthropods are identified by their morphology, that is by their outer appearance.

Moulting Process of shedding the cuticle between developmental stages (i.e. instars).

Myiasis Invasion of vertebrate organs or tissues by larvae of Diptera that feed on living or dead tissues. Myiasis may be accidental, obligatory or facultative. Accidental myiasis usually involves people eating food contaminated with fly eggs or larvae; no real harm is caused; live larvae may be passed in excreta or vomit. Obligatory myiasis is when it is essential for fly larvae to live on a live host for at least part of their life; an example is the larva of the tumbu fly. In facultative myiasis fly larvae are normally free-living, often breeding in meat or carrion, but under certain conditions they may infect living hosts; examples are larvae of bluebottles and greenbottles infesting wounds. *See* Myiasis in the Index for different forms of myiasis.

Nearctic region The USA, Canada, Greenland and northern Mexico. *See* Zoogeographical regions.

Neotropical region South America, Central America, southern Mexico and Caribbean islands. *See* Zoogeographical regions.

New World   North, Central and South America, and the Caribbean area. *See* Old World.

Nicotinoids   The nicotinoids such as imidacloprid and dinotefuran are a relatively new class of insecticides. They are related to nicotine, which has been used mainly as an agricultural insecticide. They act on the acetylcholine system, and are safe on mammals but kill a broad range of insect pests.

Nidicolous   In medical entomology usually used to describe the habit of soft, and some hard, ticks of living in and around the homes, nests, burrows, and caves of their hosts. Such ticks disperse little. *See* Non-nidicolous.

Nocturnal   Refers to activity during the night, such as blood-feeding in anophelines and appearance in vertebrate blood of nocturnally periodic microfilariae of some helminths, e.g. *Wuchereria bancrofti*.

Non-nidicolous   Describes the habits of most hard ticks of living in open and exposed habitats away from their hosts' homes. Such ticks are often dispersed by mammals and birds over considerable distances. *See* Nidicolous.

Nymph   In incomplete metamorphosis (hemimetabolous development) the stage in the life cycle that hatches from the egg (e.g. bedbugs, lice), or the stage that arises through the moulting of the larval stage (e.g. ticks).

Old World   All countries and areas east of the Americas. *See* New World.

Oocyst rate   Percentage of mosquitoes that have malarial oocysts on the stomach.

Organochlorines (chlorinated hydrocarbons)   Synthetic insecticides containing carbon, chlorine and hydrogen, e.g. DDT, dieldrin, HCH and methoxychlor.

Organophosphates   Synthetic insecticides which are derivatives of phosphoric acid and hence all contain phosphorus, e.g. diazinon, dichlorvos and malathion.

Oriental region   Asia east of Pakistan and south of the Himalayas and central China, covering Taiwan, Sri Lanka, and the South-east Asia archipelago eastwards to include Sulawesi. *See* Zoogeographical regions.

Ornithophagic (ornithophilic)   Arthropods that blood-feed on birds.

Osmoregulation   The regulation of water balance in arthropods; maintaining the homeostasis (balance) of osmotic and ionic content of the body fluids.

Overwintering   Describes the survival tactics of arthropods during winters. For instance adults (e.g. some mosquitoes) may cease feeding and ovipositing and enter a state of hibernation until warmer weather reappears and activity resumes. The growth and development of the immature stages (e.g. mosquito and simuliid larvae) may also slow down.

Palaearctic region   Europe, North Africa, Asia north of the Himalayas and central China, Japan, Iceland, mid-Atlantic islands. *See* Zoogeographical regions.

Periodicity   Several organisms, including both vectors and parasites, exhibit temporal periodicity in aspects of their behaviour. *See* Aperiodic, Diel periodicity, Diurnal, Nocturnal, Subperiodic.

Peritrophic membrane   A thin tubular sheath secreted either by cells at the anterior end of the mid-gut (mosquito larvae, tsetse-fly adults) or by cells lining the mid-gut (adult mosquitoes). It is found only in some insects. It supposedly forms a protective lining for the mid-gut, but its exact function remains largely unknown. In tsetse-flies the peritrophic membrane plays a role in the cyclical development of human sleeping sickness trypanosomes.

Pheromone   Chemical (semiochemical) released usually as an odour by an individual which produces reactions in others of the same species, e.g. sex pheromones in tsetse-flies and ticks.

Phoresy   Transport of an animal from one place to another by means of attachment to another animal, e.g. *Simulium neavei* larvae on freshwater crabs, and eggs of *Dermatobia* flies attached to adult female mosquitoes.

Polynesia   A group of numerous islands in the west Pacific extending from Hawaii to New Zealand and including the Solomon Islands, New Caledonia, Fiji, Tuvatu, Tonga, Samoa, Cook Islands, Society Islands, and Tahiti.

Polytene chromosomes   So-called giant chromosomes found only in certain tissues of Diptera, such as the ovarian nurse cells of half-gravid anophelines and larval salivary glands of simuliids. When stained they show distinct banding patterns which can often be used to identify species within species complexes. *See* Sibling species.

Promastigote (leptomonad)   Morphological form of a trypanosomatid with the flagellum arising near the anterior end, e.g. *Leishmania* parasites in the phlebotomine sand-fly gut.

Pseudopod (false leg)   Stumpy protuberances present on dipterous larvae of some species (e.g. tabanids, phlebotomine sand-flies) that assist them in locomotion. None of the larvae of the Diptera possesses true legs.

Pseudotracheae   Small tubes in the labella of the adults of some Diptera (e.g. muscids, calliphorids and tabanids) which are supported by sclerotized rings. Liquid food passes through minute openings in these pseudotracheae to the mouth of the fly.

Puparium   Life-stage resulting from the hardening and sclerotization of the cuticle of the last larval instar of certain Diptera, e.g. tsetse-flies and house-flies. Equivalent to the pupa of other insects.

Pyrethroids (synthetic pyrethroids)   Synthetic insecticides containing different pyrethrin-like chemicals, e.g. permethrin, deltamethrin, lambda-cyhalothrin.

Questing   The behaviour of ticks, mainly ixodids, when climbing up vegetation, such as grasses and herbaceous plants, in order to actively seek out passing hosts, to which they attach themselves.

Quiescence   Temporary state of arrested or slowed development, such as in ixodid larvae after blood-feeding but prior to moulting to the nymphal stage, or state of inactivity of some hibernating adult mosquitoes.

Reservoir host   An animal in which populations of disease organisms persist indefinitely, and which passes the disease to other species of hosts, often by vectors. Reservoir hosts may be maintenance hosts, and are often mammals and birds. Some arthropod vectors which are long-lived, and may also be capable of transovarial transmission, are sometimes regarded as reservoir hosts, e.g. ticks as vectors of relapsing fever and various viruses.

Residual spraying   Application of a persistent insecticide (e.g. malathion) to surfaces such as to the inside walls and roofs of houses in malaria control programmes, and to trees in tsetse control projects.

Rickettsiae   A group of Gram-negative intracellular coccoid-shaped bacteria, many of which are transmitted by arthropods, e.g. *Rickettsia prowazekii* transmitted by body lice and causing typhus. Formerly rickettsiae were regarded as micro-organisms intermediate between bacteria and viruses.

Seed tick   Name often given to the very small larva of an ixodid tick before it has blood-fed.

Sclerotization   Process which results in the new arthropod cuticle formed after moulting being tanned (darkened) and hardened to give it rigidity.

Sibling species (isomorphic or cryptic species)   Species that are morphologically indistinguishable or almost so, and which are recognized by non-morphological features, usually their polytene chromosomes (e.g. species of the *Simulium damnosum* complex). In nature they are reproductively isolated from each other. Sibling species may differ only slightly biologically but be significantly different in epidemiological importance.

s. l. (sensu lato)   Means in the broad sense. When placed after the name of a species denotes that reference is made not just to that species but also to closely related species (e.g. sibling species) within a complex. *See* s. str.

Source reduction   Simple measures that either prevent breeding of arthropods or eliminate their breeding sites. Mainly applicable to mosquito control, e.g. covering water tanks, filling in puddles or removing discarded water-retaining receptacles.

Species   A group of individuals in natural populations that can interbreed by mating within the group and producing fertile progeny; individuals are usually similar in appearance and behaviour.

Species complex   A group of sibling, or very closely related, species that are morphologically indistinguishable (isomorphic) but are reproductively isolated, and which often live in the same area (sympatric). Species within a complex can often be identified by their polytene chromosomes or by biochemical or molecular techniques. The Diptera contain many complexes, such as in mosquitoes (e.g. *Anopheles gambiae* complex) and simuliids (e.g. *Simulium damnosum* complex).

Species group   Used for an assemblage of closely related species that although they may be morphologically similar in one or more life-stages can nevertheless be distinguished on external appearance as distinct species, e.g. the *Simulium neavei* group – comprising eight species including *S. neavei*, *S. woodi* and *S. nyasalandicum*.

Species sanitation   Control, or eradication, directed against just one species of vector or pest in a particular area, usually using simple techniques.

Spermatheca   One or more receptacles in the female reproductive system of arthropods which receive sperm during mating.

Spirochaetes   Gram-negative bacteria which have a more or less spiral shape, e.g. *Borrelia duttonii*, which is spread by soft ticks and causes relapsing fever.

Sporogony   That part of the sexual cycle of sporozoans (e.g. *Plasmodium* species) in which sporozoites are produced.

Sporozoite rate   Percentage of mosquitoes with malarial sporozoites in their salivary glands.

s. str. (sensu stricto)   Means in the strict, or narrow, sense. When placed after a species name denotes that only that species is being referred to, and not any of its closely related (e.g. sibling) species in a complex. Sometimes abbreviated to s. s. *See* s. l.

Sterile male release (sterile-insect technique – SIT)   In genetic control programmes the inundative release of large numbers of artificially sterilized male arthropods into field populations in the hope that this will result in sterile matings and consequently a reduction in population size.

Subperiodic   As applied to microfilariae in peripheral vertebrate blood, means they exhibit partial diel (24-hour) periodicity. That is, their concentration in the blood decreases from a maximum to a minimum, but not close to zero as with microfilariae showing periodic periodicity. Diurnal subperiodicity occurs in the Pacific strain of *Wuchereria bancrofti* while nocturnal subperiodicity is found in populations of *Brugia malayi* in West Malaysia, Thailand, etc.

Subspecies   In zoology the only taxon recognized below the rank of species. Subspecies differ morphologically from other members of the species and are spatially isolated, e.g. either geographically or by their hosts. Consequently they are normally reproductively isolated, but when brought together can interbreed with other subspecies. Subspecies may eventually evolve into distinct species.

Sylvatic   In epidemiology means that diseases are contracted in woods or forests, e.g. the forest cycle of yellow fever. *See* Campestral.

Synanthropic   Applied to animals living in close association with humans or their houses, e.g. house-flies and triatomine bugs.

Synergist   Chemical that has little or no toxicity but, when combined with some insecticides, enhances their activity and thus reduces dosage rates. For example, piperonyl butoxide is a synergist added to the insecticide pyrethrum.

Transovarial transmission   Production by an infected vector of eggs infected with parasites (e.g. viruses, rickettsiae). When they hatch they give rise to individuals (e.g. larvae, nymphs) that are infected and are either capable of transmitting the parasites (e.g. ticks) or passing it on to later life-cycle stages that transmit the infection (e.g. scrub typhus mites). *See* Hereditary transmission.

Transstadial transmission   Survival of parasites through successive life-cycle stages (larva → nymph(s) → adult) of an organism (e.g. ticks), each of which can transmit the parasite if it is haematophagous.

Trypomastigote   Morphological form of a trypanosome with the flagellum arising near the posterior end, and running the length of the body where it is attached to the cell membrane. Trypomastigotes are found in the vertebrate blood of hosts infected with trypanosomiasis, and are the form ingested by a vector with its blood-meal.

Ultra-low-volume (ULV)   Refers to spraying an insecticide in concentrated form; thus the dosage rate is small, e.g. <5 litres/hectare.

Vector   Organism that conveys an aetiological agent from one host to another. A vector may be an intermediate host (e.g. culicines transmitting filariasis) or not (e.g. house-flies mechanically transmitting bacteria).

Venereal transmission   When pathogens, such as viruses and rickettsia, are passed from congenitally infected males to females during mating, Occurs in some vectors of yellow fever and Crimean–Congo haemorrhagic fever.

Viraemia   Presence of virus in vertebrate blood. High viraemias are usually necessary for transmission by arthropod vectors.

Wettable powder (water-dispersible powder)   Technical grade insecticide diluted with an inert carrier (dust) and to which a wetting agent or surfactant has been added. The resultant wettable powder is then mixed with water for spraying onto surfaces.

Zoogeographical regions   Any one of the six main geographical areas referred to by zoologists. Each region has its own particular fauna, of which many species occur only in that region. The Palaearctic and Nearctic region are sometimes grouped together and referred to as the

Holarctic region. *See* Afrotropical, Australasian, Nearctic, Neotropical, Oriental and Palaearctic regions.

Zoonosis Natural transmission of infections between vertebrate hosts and humans.

Zoophagic (zoophilic) Blood-sucking arthropods that feed on non-human animals.

Select bibliography

Centers for Disease Control and Prevention (2008) *Travelers' Health: Yellow Book. Health Information for International Travel.* Atlanta, GA: CDC. www.cdc.gov/travel/ybToc.aspx.

Clark, G. G. (coordinator) (1994) Prevention of tropical diseases: status of new and emerging vector control strategies. Proceedings of a symposium on vector control. *American Journal of Tropical Medicine and Hygiene,* **50** (6, suppl.), 1–159.

Eldridge, B. F. and Edman, J. D. (eds.) (2000) *Medical Entomology: a Textbook on Public Health and Veterinary Problems Caused by Arthropods.* Dordrecht: Kluwer.

Gratz, N. (2006) *Vector- and Rodent-borne Diseases in Europe and North America: Distribution, Public Health Burden and Control.* Cambridge: Cambridge University Press.

Kettle, D. S. (1995) *Medical and Veterinary Entomology,* 2nd edn. Wallingford: CAB International.

Lane, R. P. and Crosskey, R. W. (eds.) (1993) *Medical Insects and Arachnids.* London: Chapman and Hall.

Lehane, M. (2005) *The Biology of Blood-Sucking in Insects,* 2nd edn. Cambridge: Cambridge University Press.

Marquardt, W. C. (ed.) (1996) *Biology of Disease Vectors,* 2nd edn. Amsterdam: Elsevier Academic Press.

Mullen, G. and Durden, L. (eds.) (2002) *Medical and Veterinary Entomology.* Amsterdam: Academic Press.

Service, M. W. (ed.) (2001) *The Encyclopedia of Arthropod-Transmitted Infections of Man and Domesticated Animals.* Wallingford: CABI.

van Emden, H. F. and Service, M. W. (2004) *Pest and Vector Control.* Cambridge: Cambridge University Press.

World Health Organization (1997) *Vector Control: Methods for Use by Individuals and Communities,* prepared by J. A. Rozendaal. Geneva: WHO.

World Health Organization (2006) *Pesticides and their Appplication. For the Control of Vectors and Pests of Public Health Importance,* 6th edition. WHO/CDS/NTD/WHOPES/GCDPP. Geneva: WHO.

World Health Organization (2008) *Public Health Significance of Urban Pests,* ed. X. Bonnefoy, H. Kampen and K. Sweeney. Copenhagen: WHO.

Index